Count & Write

Number Tracing and Activity Workbook

This book belongs to:

Want freebies?

Email us at
hello@edventurelearning.com
with the subject "Count and Write"
and we'll send you free printables to
keep the learning going!

Animal illustrations by Michael Rayback Design

Printed in the United States of America
ISBN: 978-1-64824-022-5

www.edventurelearning.com
Questions? Email us at hello@edventurelearning.com

1 - 10

Sheep is knitting a scarf.

How many balls of yarn are there?

1

one

There is one ball of yarn.

Draw 1 circle.

Color in 1 square.

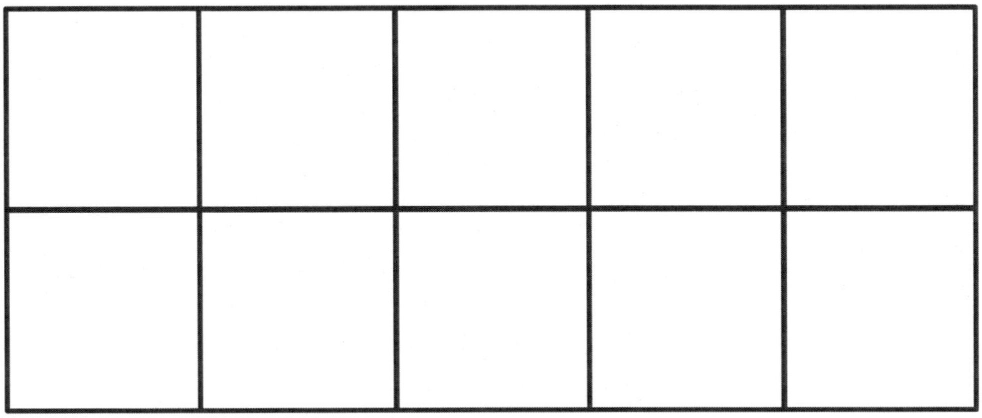

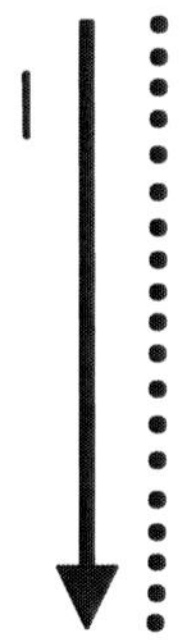

one

one one

one one

one one

one

COLOR AND COUNT

Donkey is picking vegetables.

How many baskets of vegetables does he have?

2

two

There are two baskets.

Draw 2 circles.

Color in 2 squares.

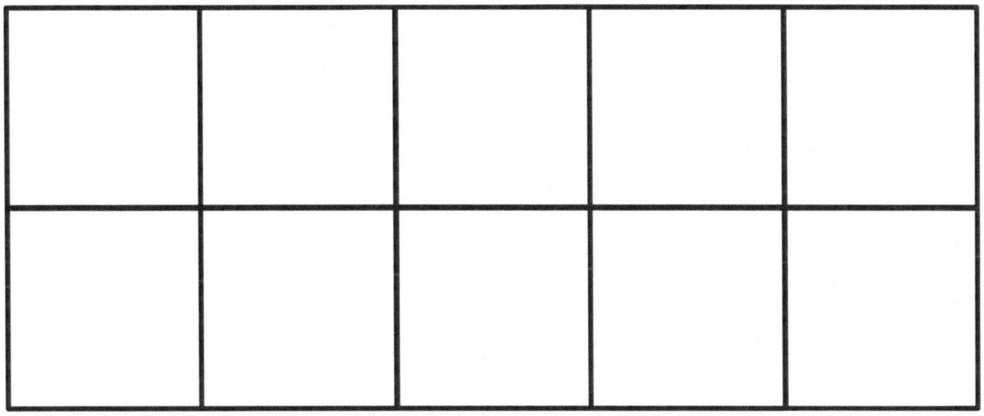

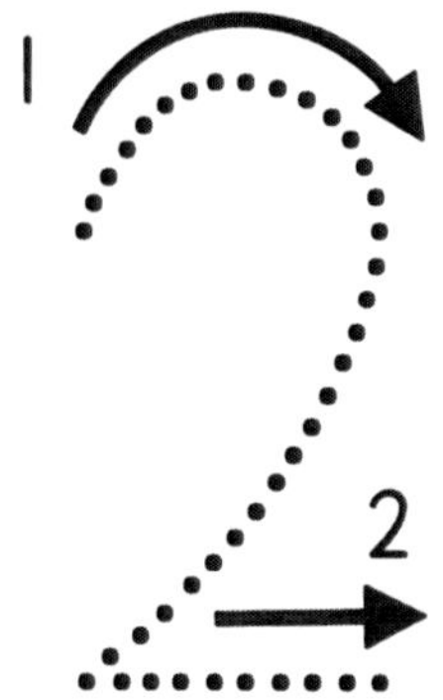
1
2

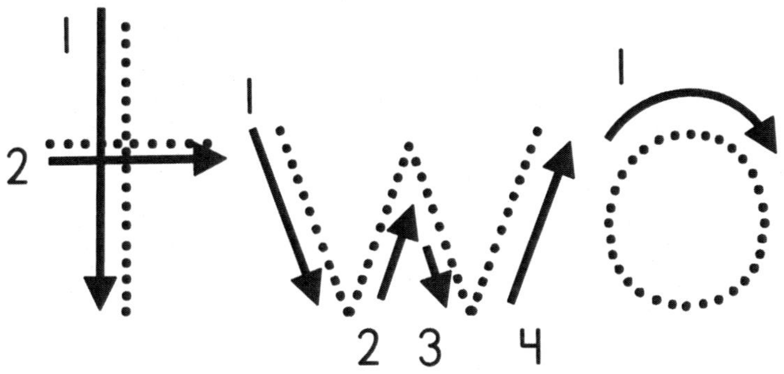

two two

two two

two two

two

COLOR AND COUNT

Raccoon is balancing books.

How many books are on her head?

3
three

There are three books.

Draw 3 circles.

Color in 3 squares.

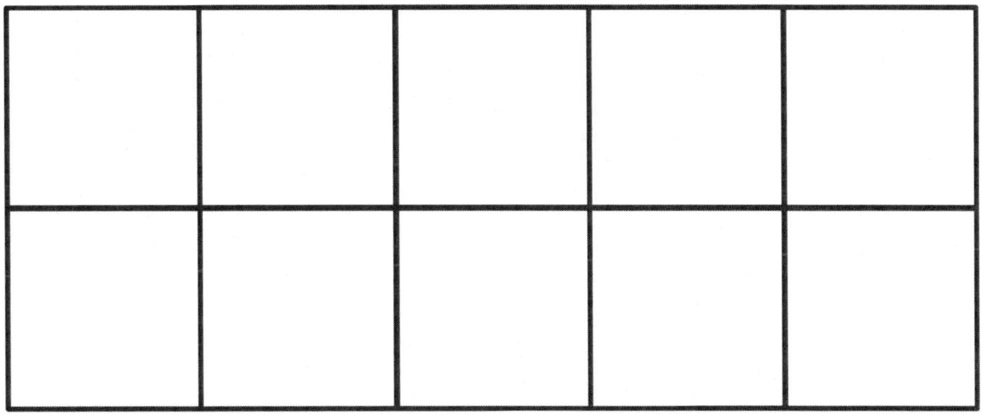

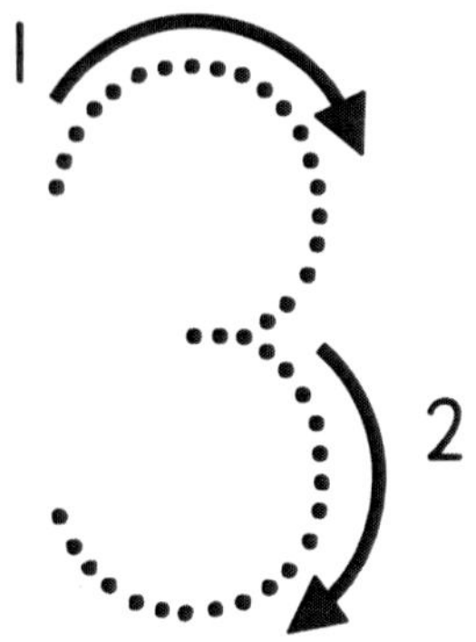

3

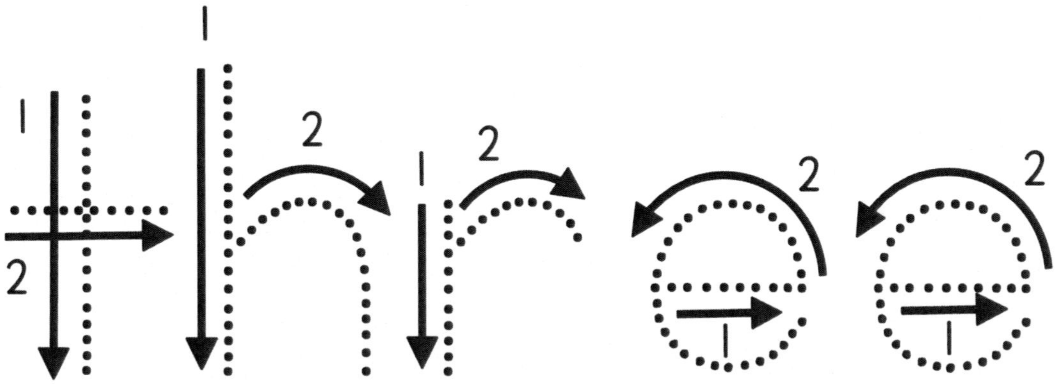

three three

three three

three three

three

COLOR AND COUNT

Daddy and Mommy Cat are getting their kittens ready for bed.

How many kittens do they have?

4
four

They have four kittens.

Draw 4 circles.

Color in 4 squares.

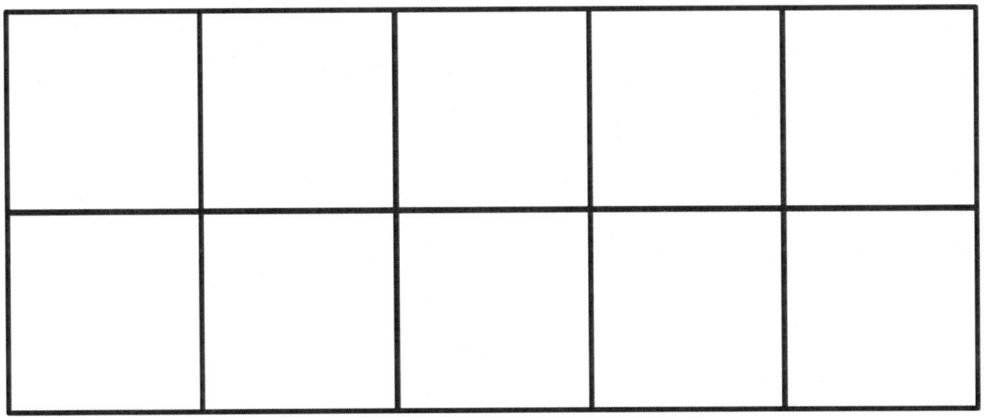

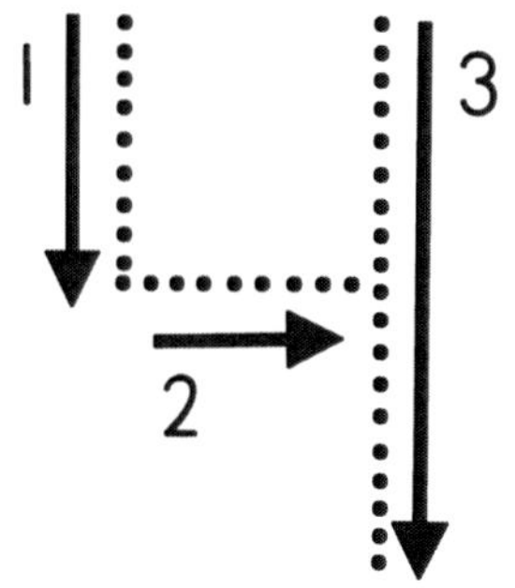
1
2
3

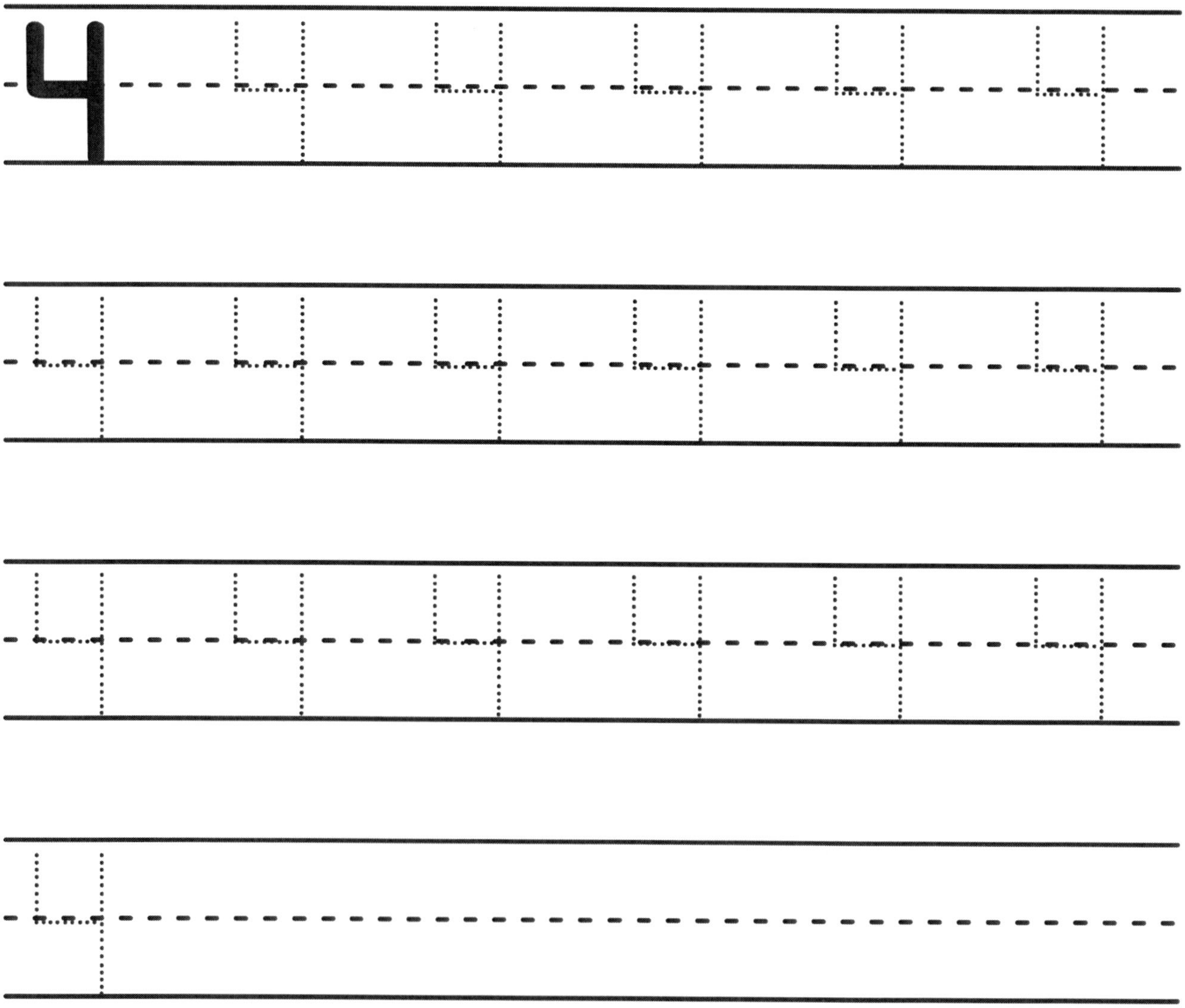

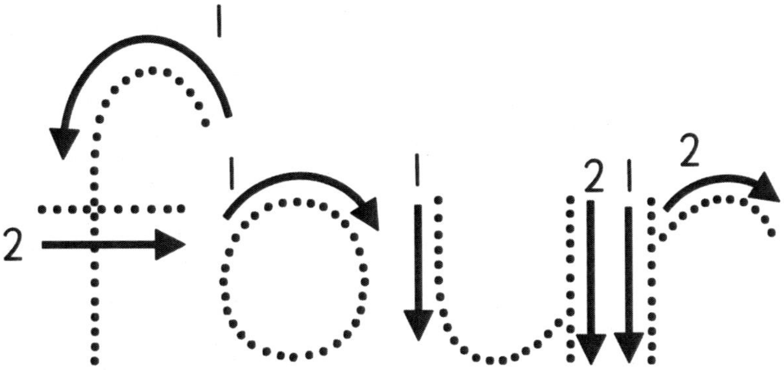

four four

four four

four four

four

COLOR AND COUNT

Stork is flying paper airplanes.

How many paper airplanes does he have?

5
five

There are five paper airplanes.

Draw 5 circles.

Color in 5 squares.

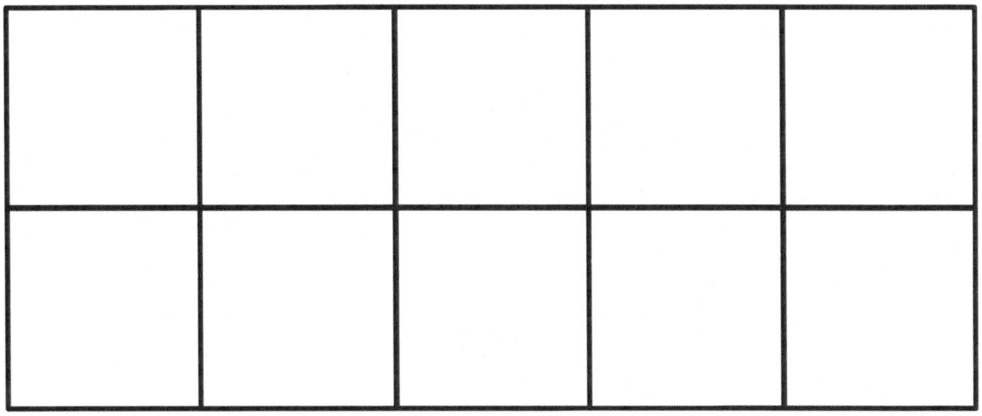

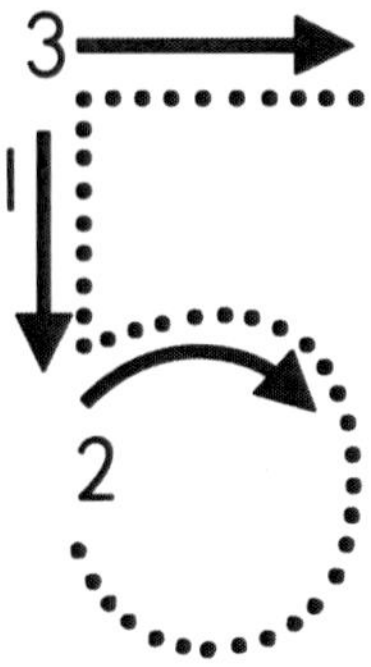
3
1
2

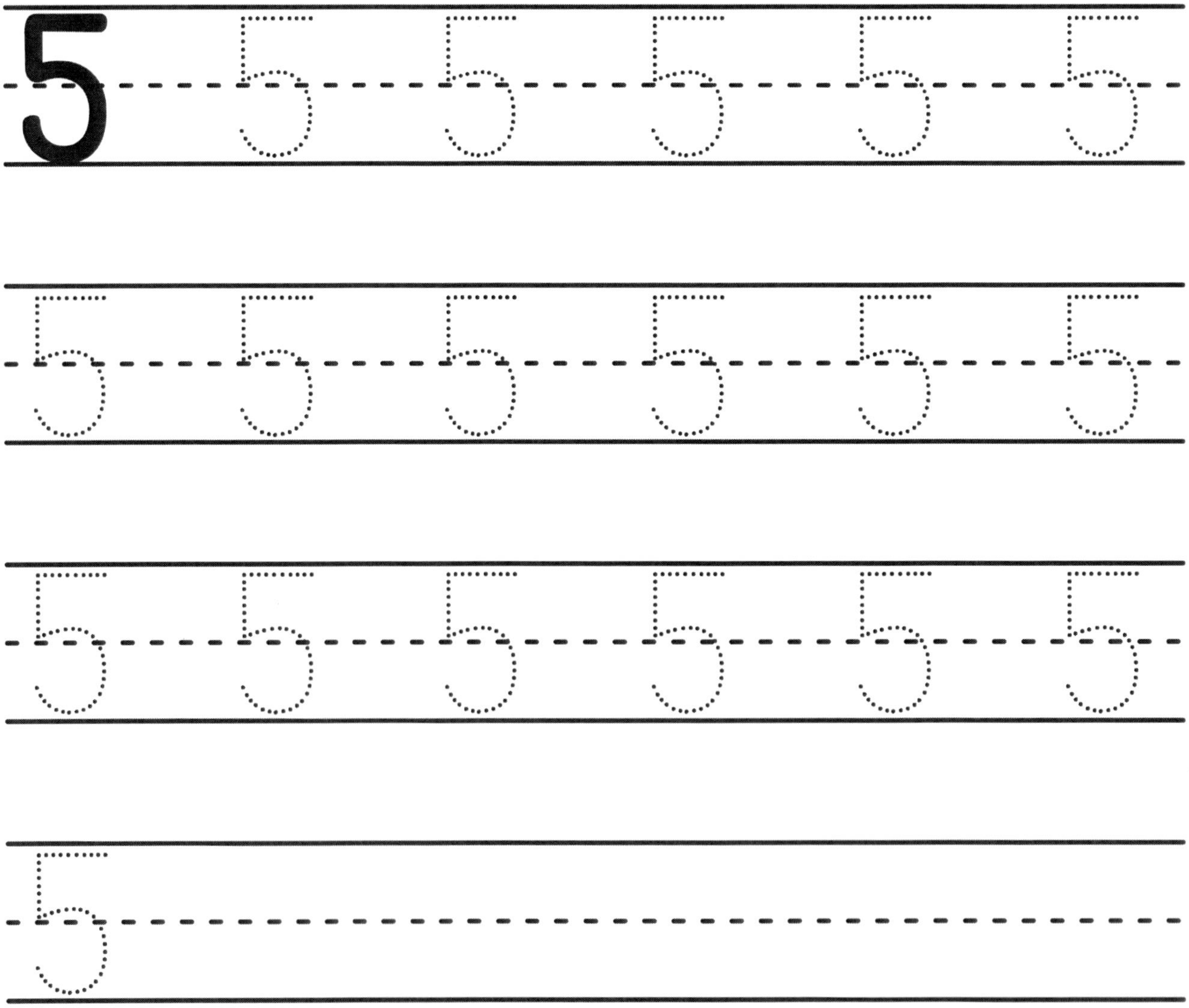
5

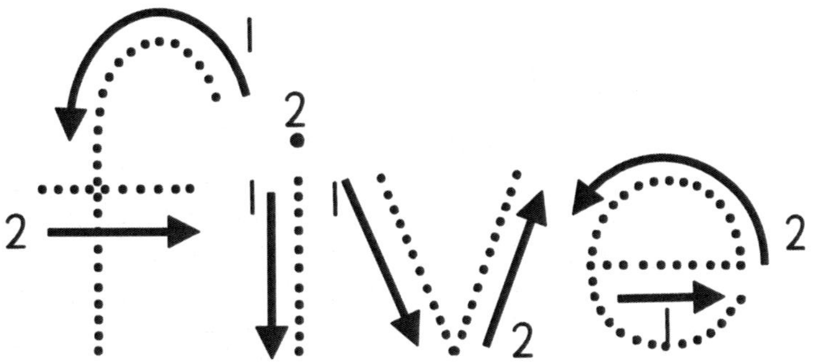

five five

five five

five five

five

COLOR AND COUNT

Lion is juggling.

How many balls is he juggling?

6

six

There are six balls.

Draw 6 circles.

Color in 6 squares.

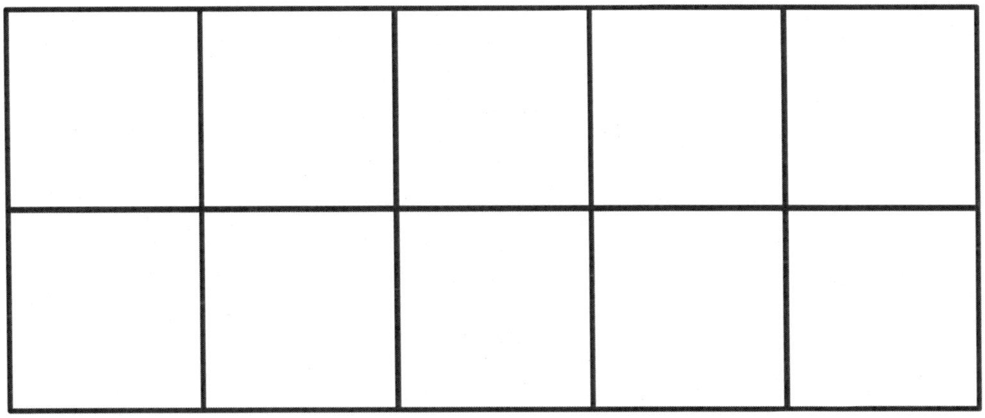

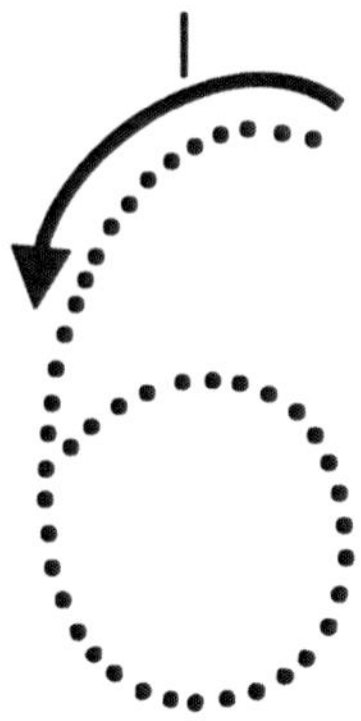

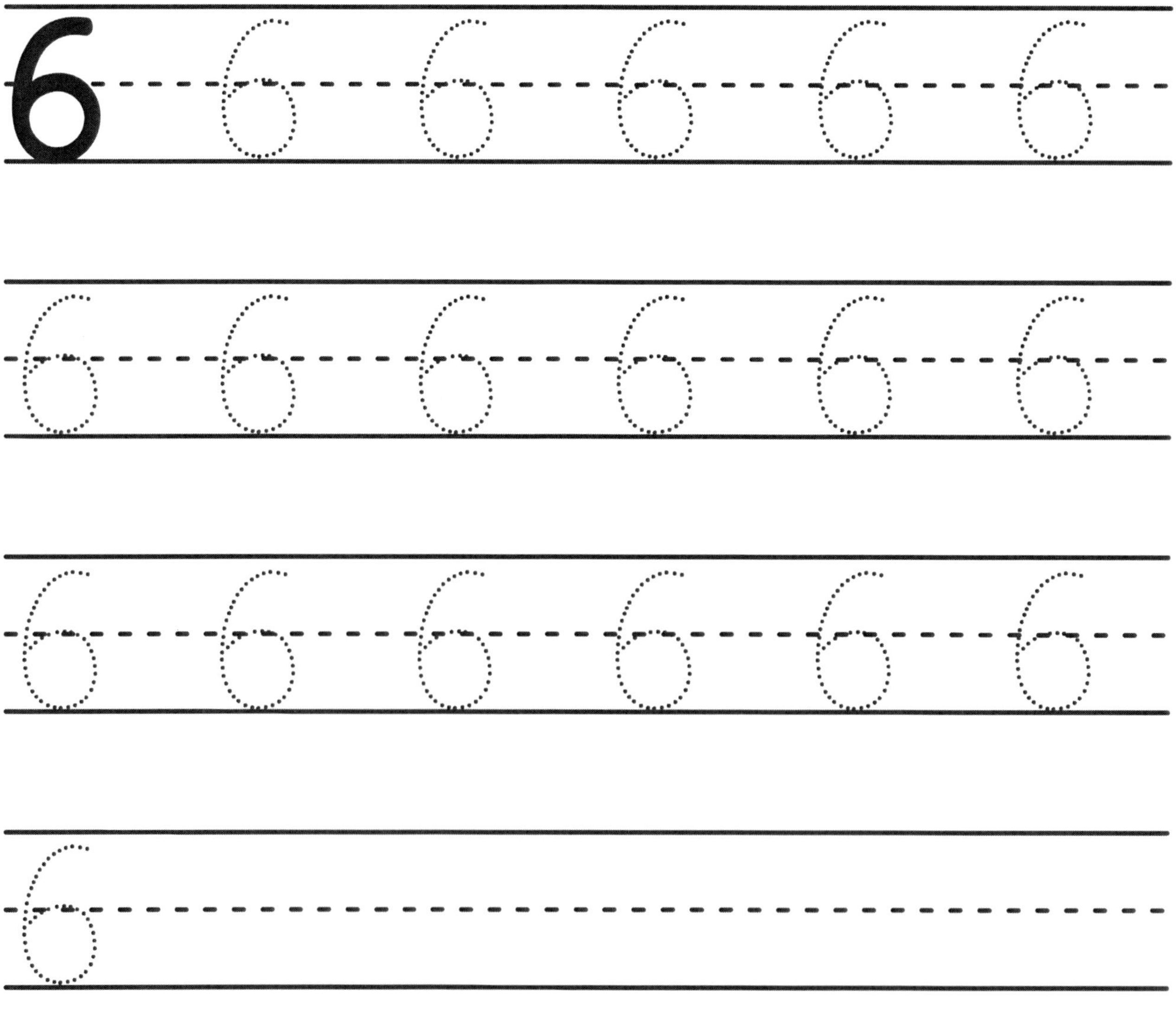

six

six six

six six

six six

six

COLOR AND COUNT

Penguin is skiing.

How many snowflakes are in the air?

7

seven

There are seven snowflakes.

Draw 7 circles.

Color in 7 squares.

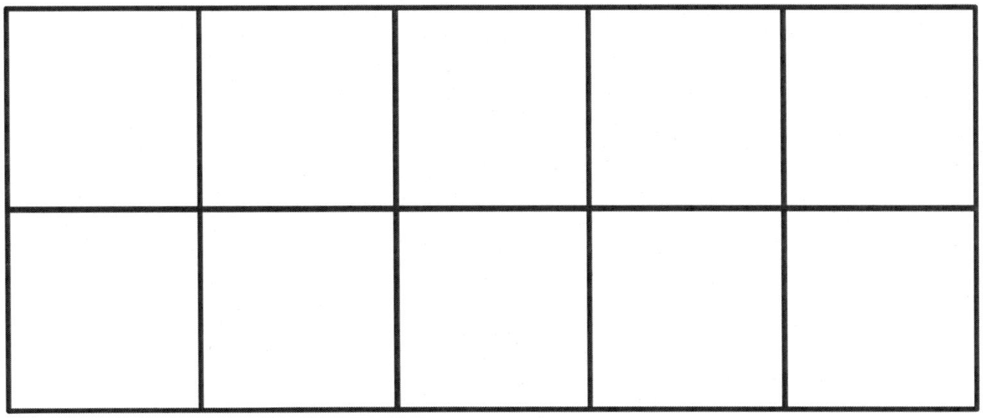

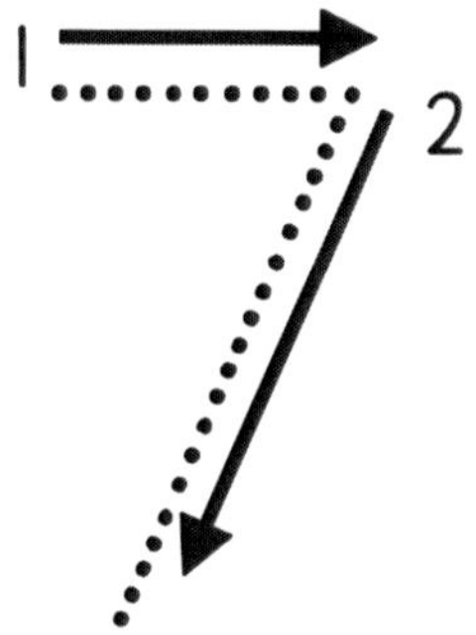
1
2

7

seven

seven seven

seven seven

seven seven

seven

COLOR AND COUNT

Fox is playing the flute.

How many music notes are there?

8
eight

There are eight music notes.

Draw 8 circles.

Color in 8 squares.

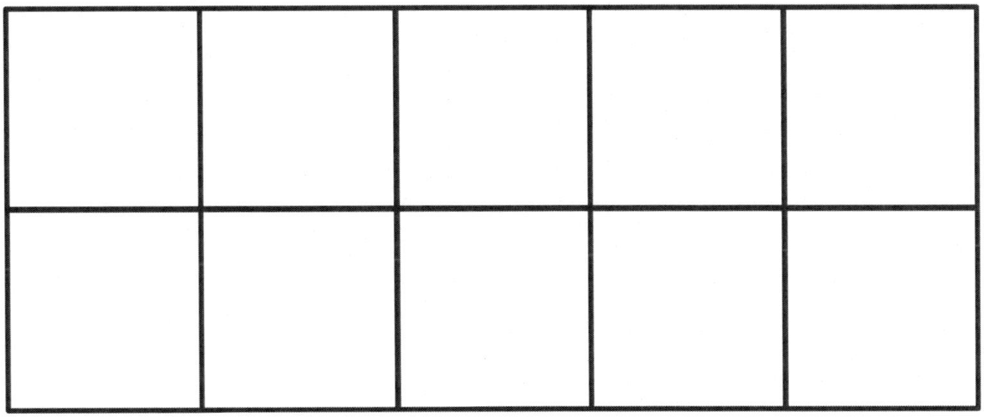

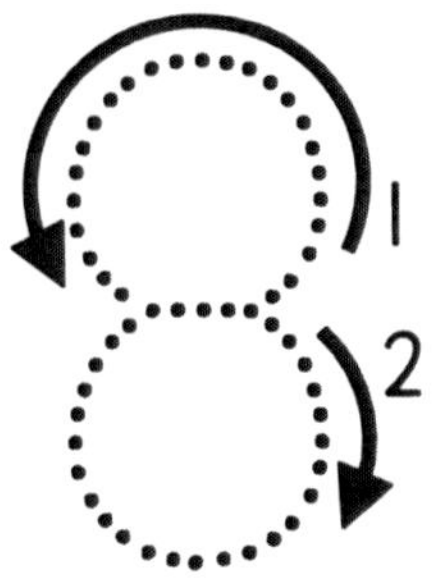

8

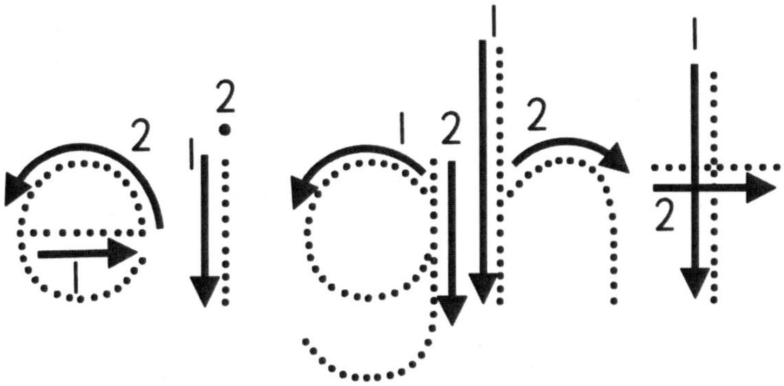

eight eight

eight eight

eight eight

eight

COLOR AND COUNT

Pig is working in the garden.

How many carrots are growing in the ground?

9

nine

There are nine carrots.

Draw 9 circles.

Color in 9 squares.

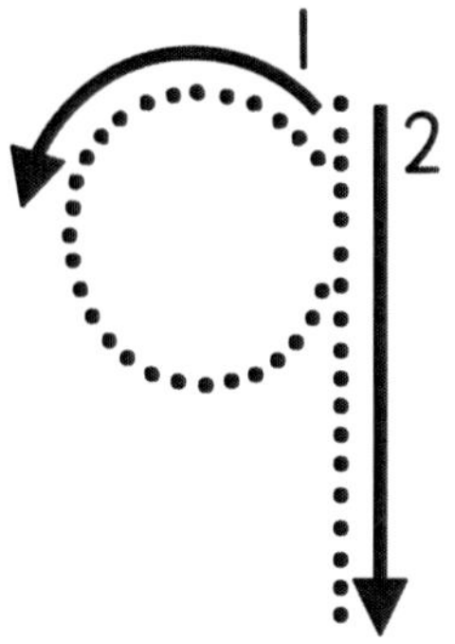

q

nine

nine nine

nine nine

nine nine

nine

COLOR AND COUNT

Beaver is building a lodge.

How many logs does he have?

10

ten

There are ten logs.

Draw 10 circles.

Color in 10 squares.

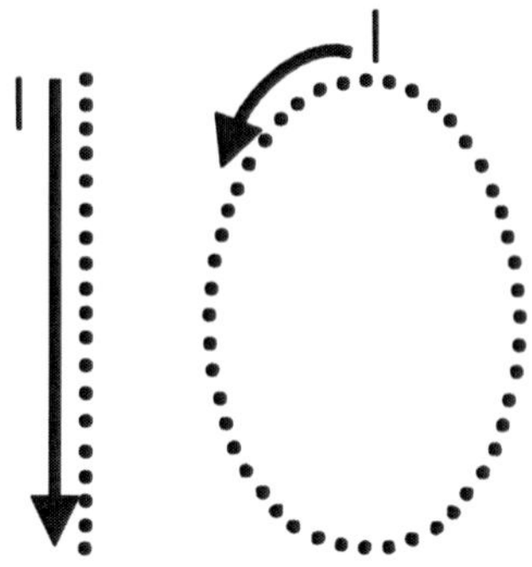

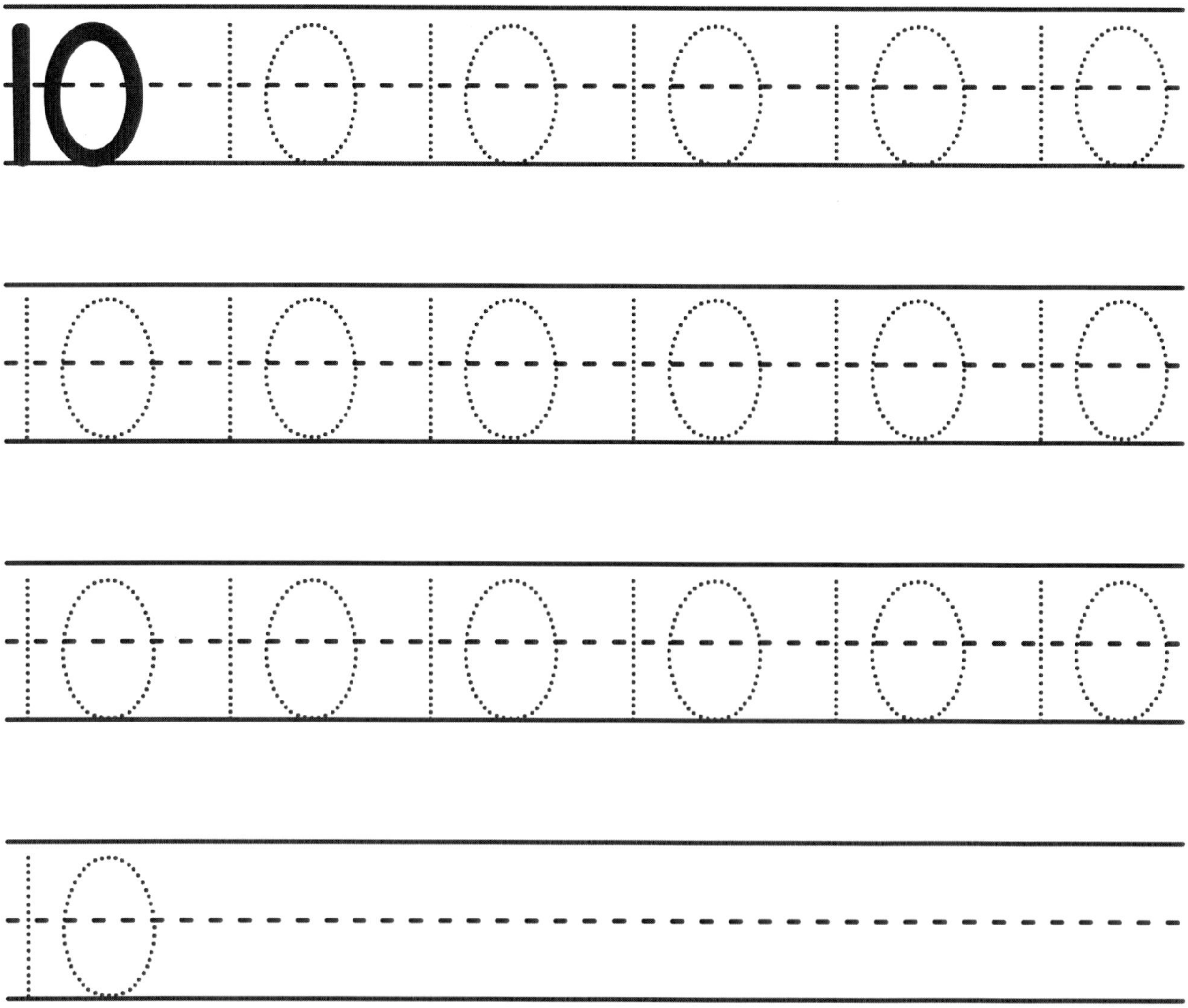

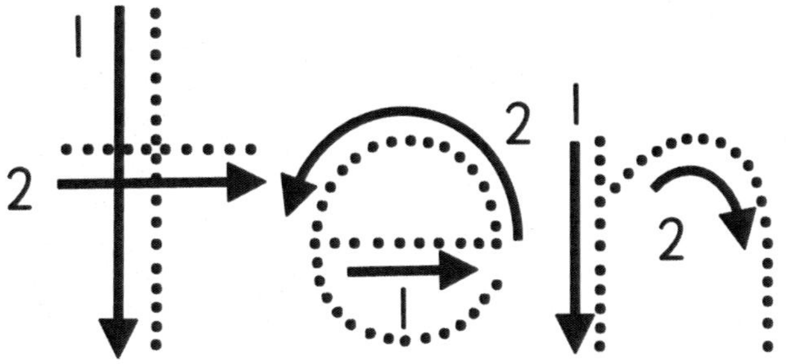

ten ten

ten ten

ten ten

ten

TRACE AND WRITE

1

2

3

4

5

TRACE AND WRITE

6

7

8

9

10

HOW MANY SHAPES?

HOW MANY SHAPES?

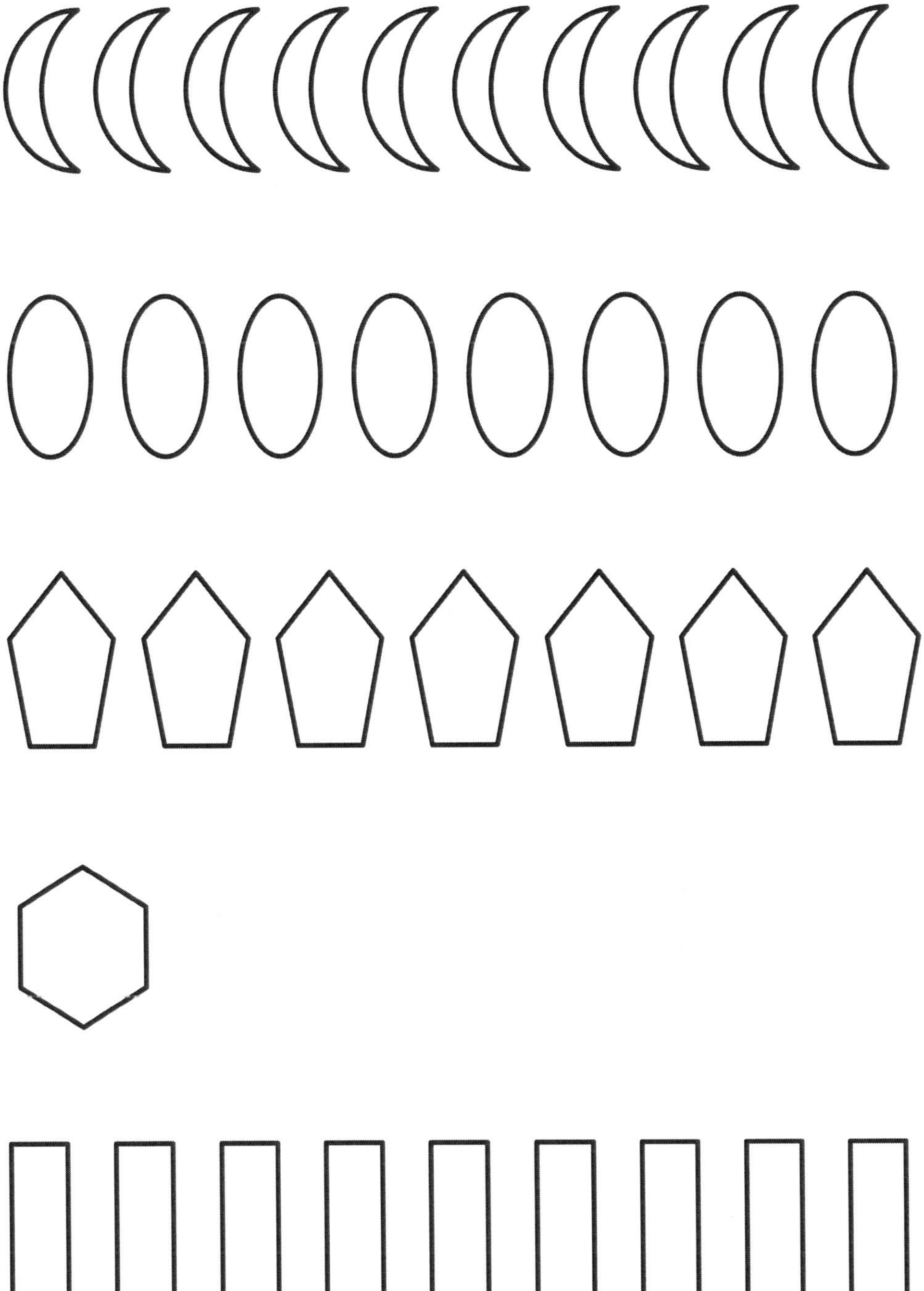

DRAW I HORN TO TURN THE HORSE INTO A UNICORN

DRAW 2 GLOVES ON THE LIZARD

ADD 3 SCOOPS OF ICE CREAM TO THE CONE

DRAW 4 EGGS IN THE NEST

DRAW 5 PIECES OF CHEESE ON THE PLATE

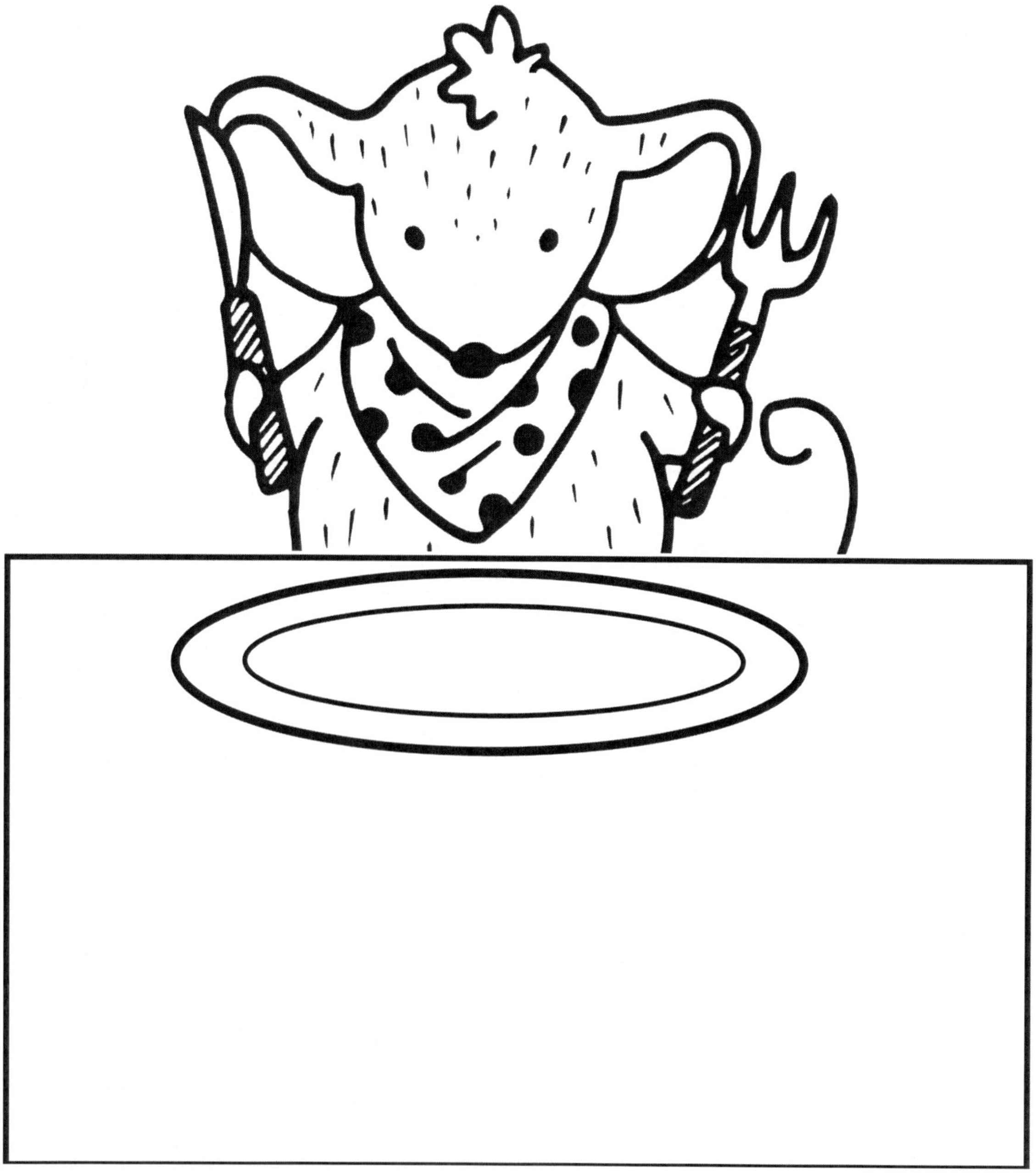

DRAW 6 WORMS IN THE BUCKET

ADD 7 STARS TO THE NIGHT SKY

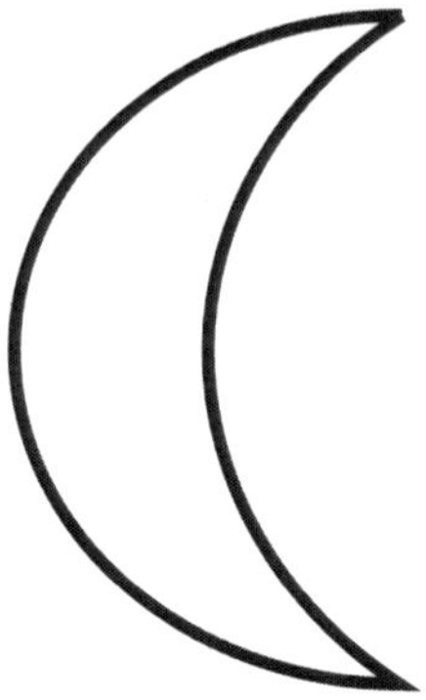

DRAW 8 RAYS ON THE SUN

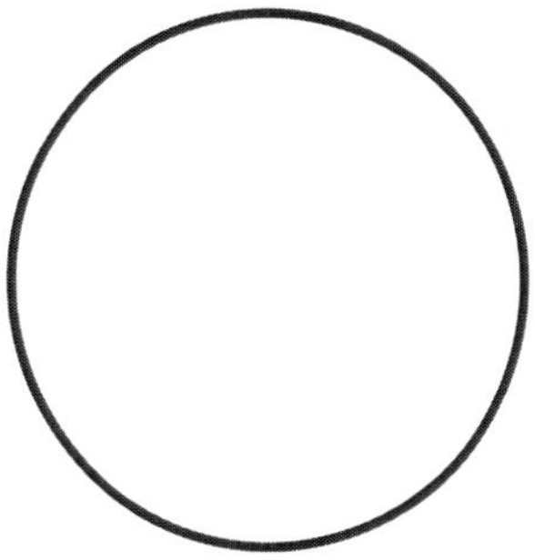

ADD 9 TAIL FEATHERS TO THE TURKEY

DRAW 10 RAINDROPS

11-20

11

eleven

11

12

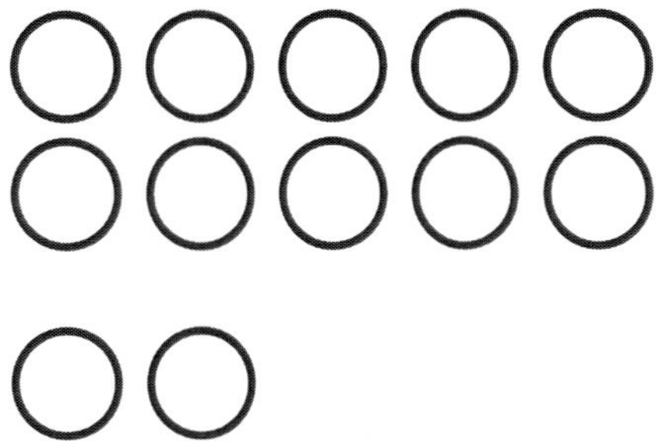

twelve

12 12 12 12 12

12 12 12 12 12

12 12 12 12 12

12

13

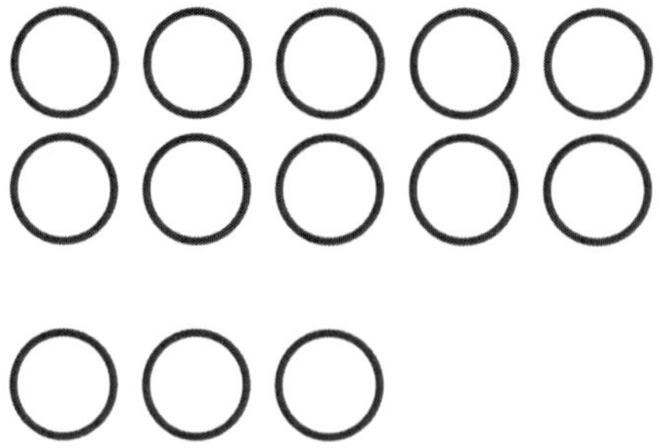

thirteen

13 13 13 13 13

13 13 13 13 13

13 13 13 13 13

13

14

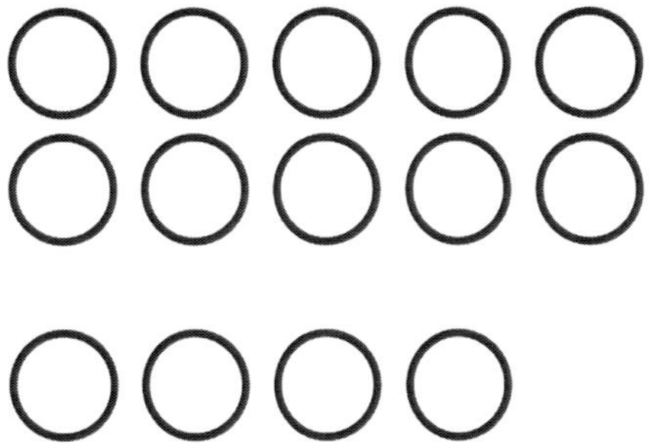

fourteen

15

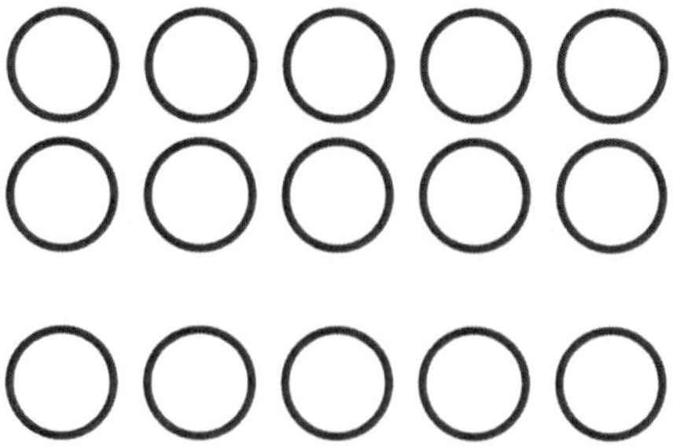

fifteen

15 15 15 15 15

15 15 15 15 15

15 15 15 15 15

15

16

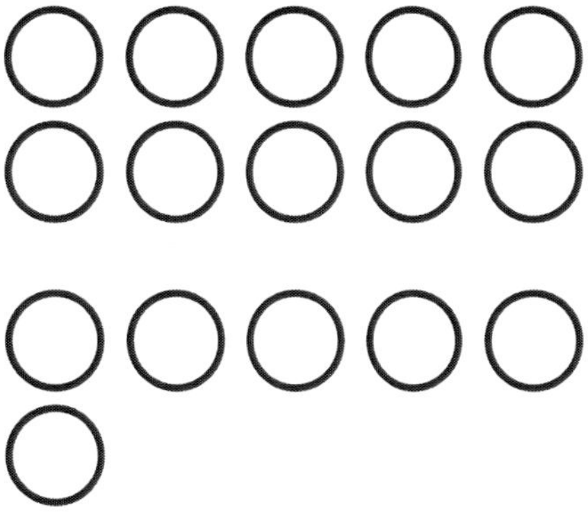

sixteen

16 16 16 16 16

16 16 16 16 16

16 16 16 16 16

16

17

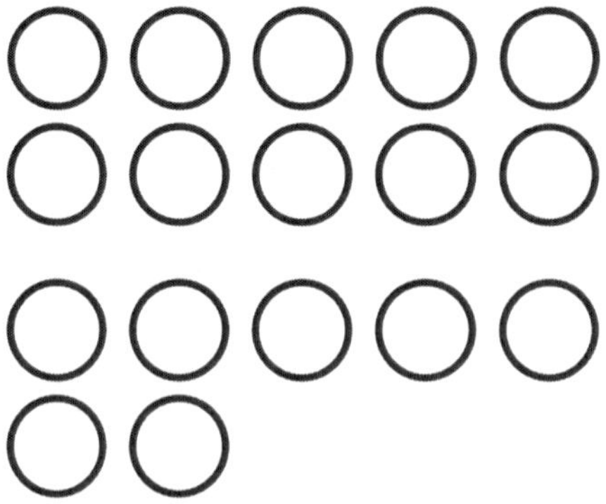

seventeen

17

18

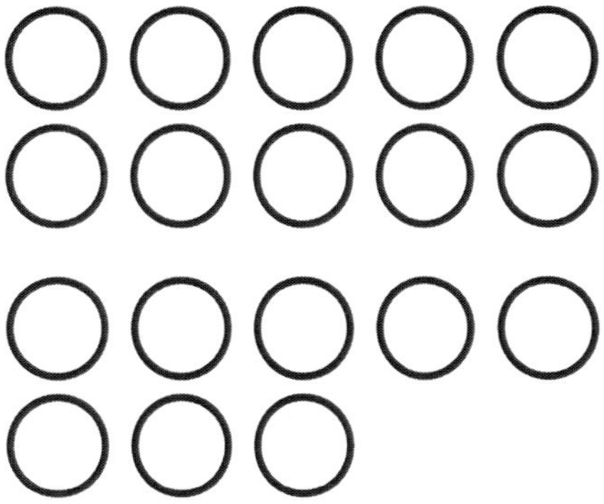

eighteen

18 18 18 18 18

18 18 18 18 18

18 18 18 18 18

18

19

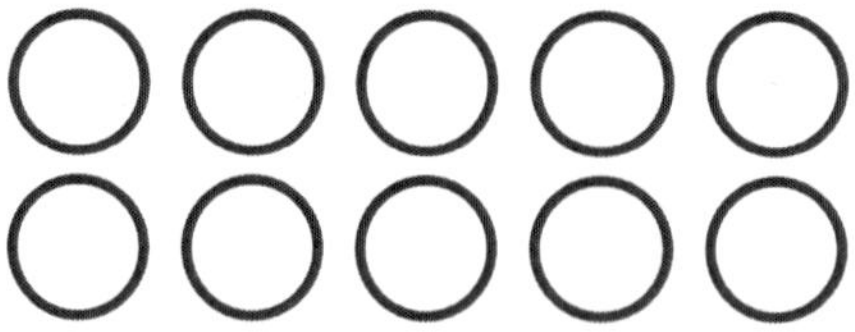

nineteen

19

20

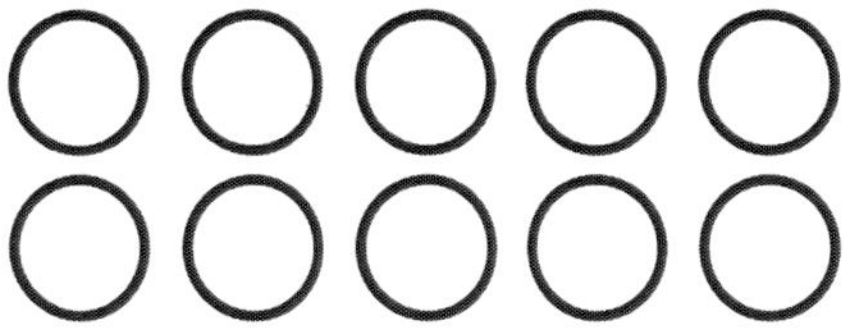

twenty

20 20 20 20 20 20 20 20 20 20

20 20 20 20 20 20 20 20 20 20

20 20 20 20 20 20 20 20 20 20

20

TRACE AND WRITE

11

12

13

14

15

TRACE AND WRITE

16

17

18

19

20

21-100

21

twenty-one

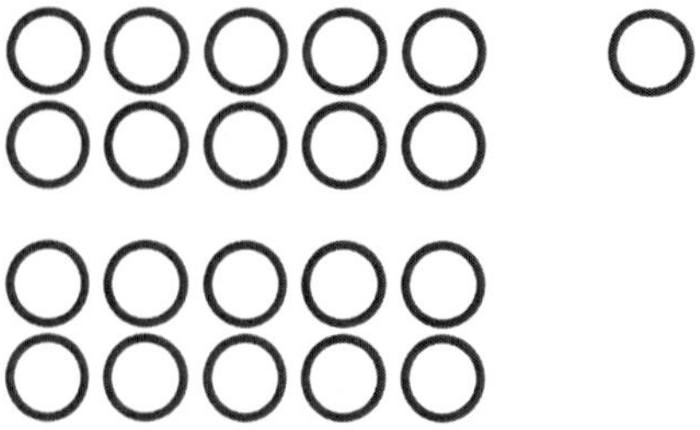

22

twenty-two

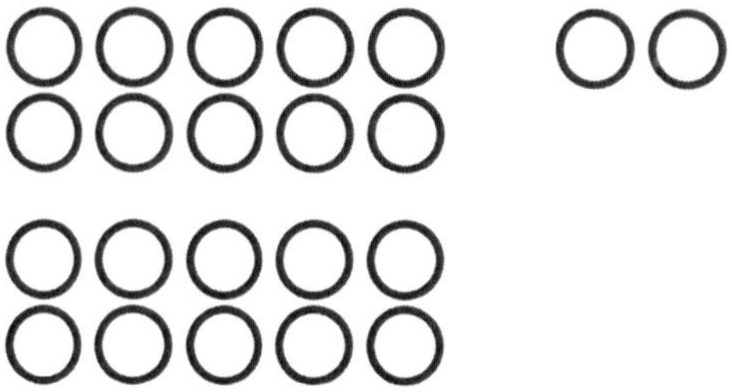

23

twenty-three

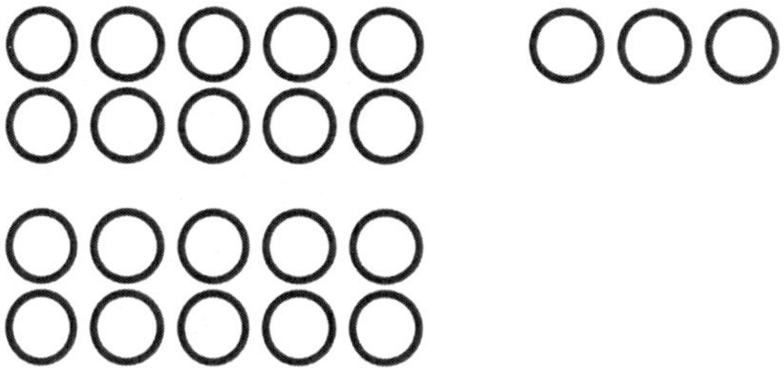

24

twenty-four

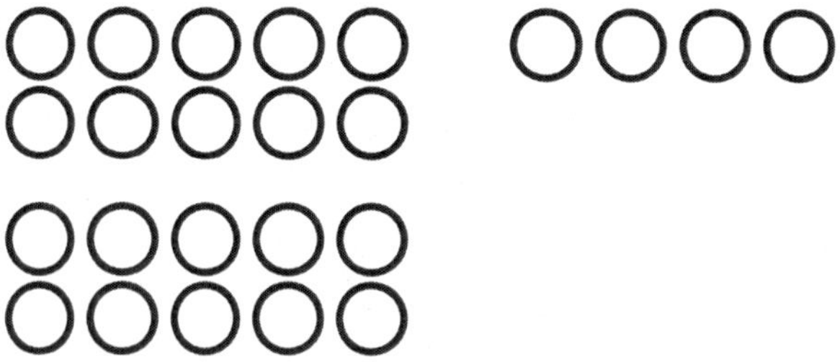

25

twenty-five

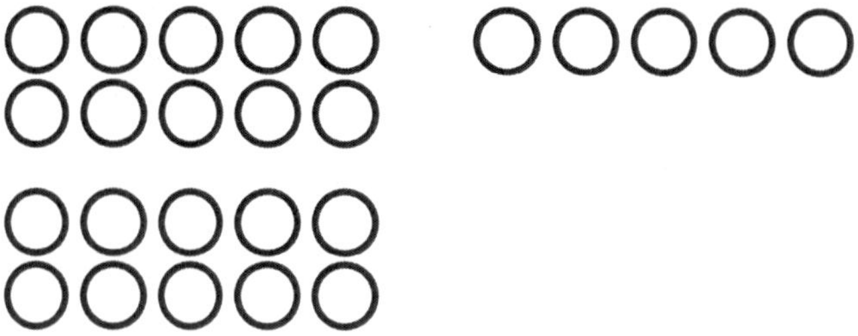

26
twenty-six

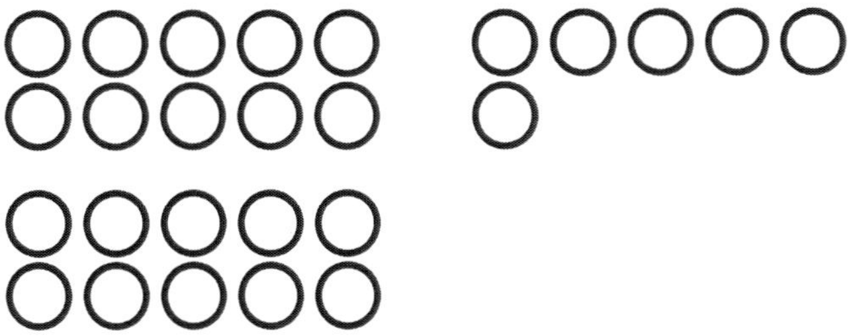

26

27
twenty-seven

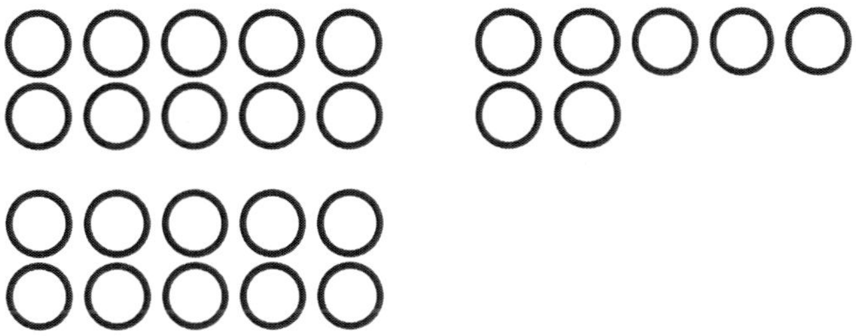

27

28
twenty-eight

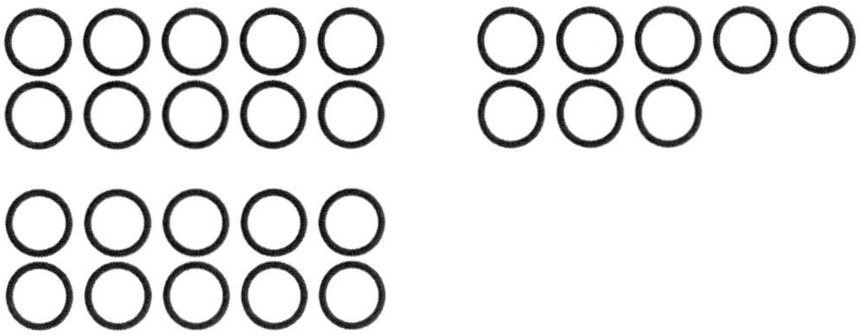

28

29
twenty-nine

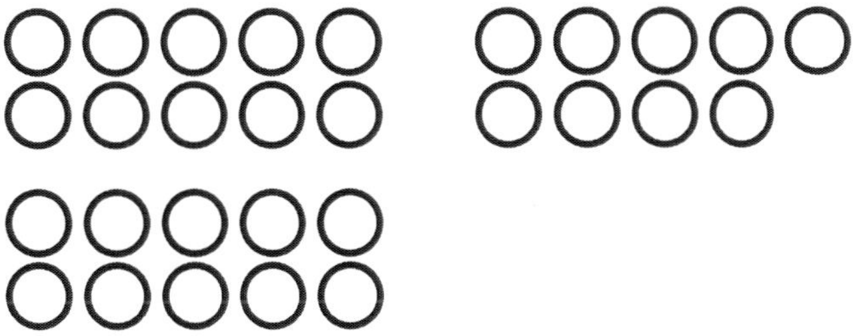

29

30
thirty

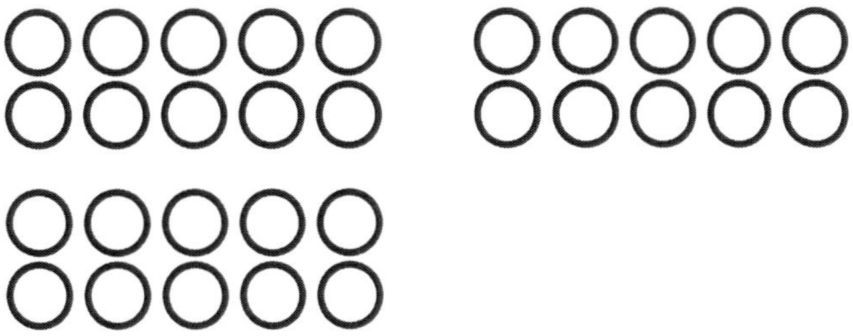

30

31

thirty-one

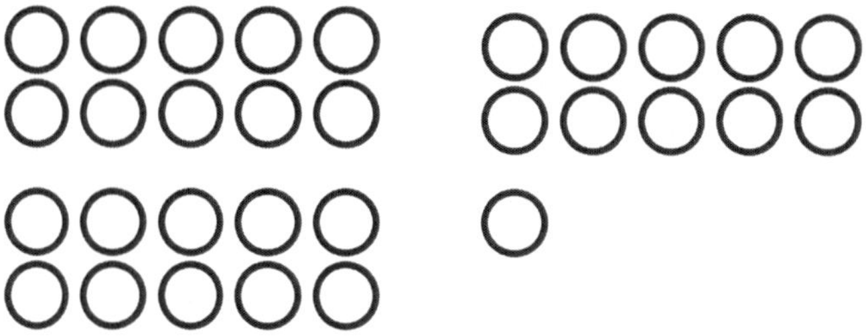

31

32

thirty-two

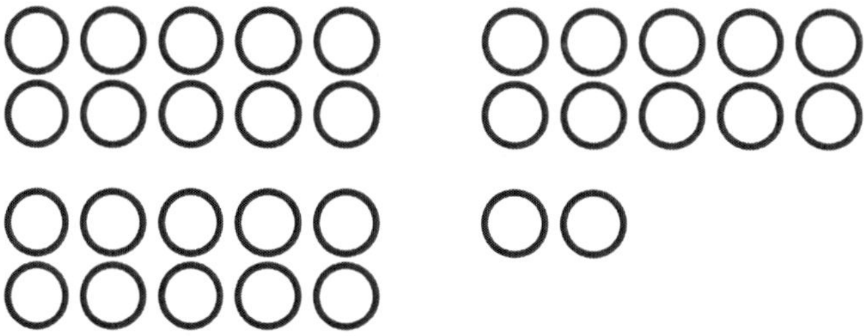

32

33

thirty-three

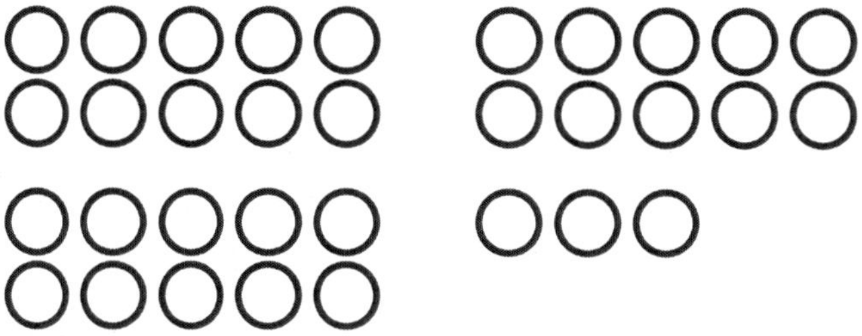

33

34

thirty-four

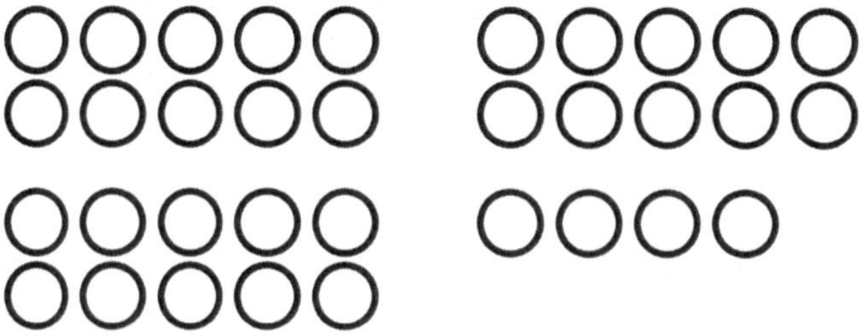

34

35

thirty-five

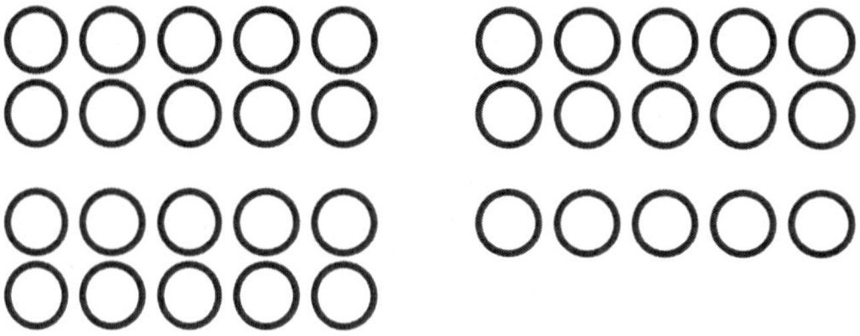

35

36

thirty-six

37

thirty-seven

38

thirty-eight

39

thirty-nine

40

forty

41
forty-one

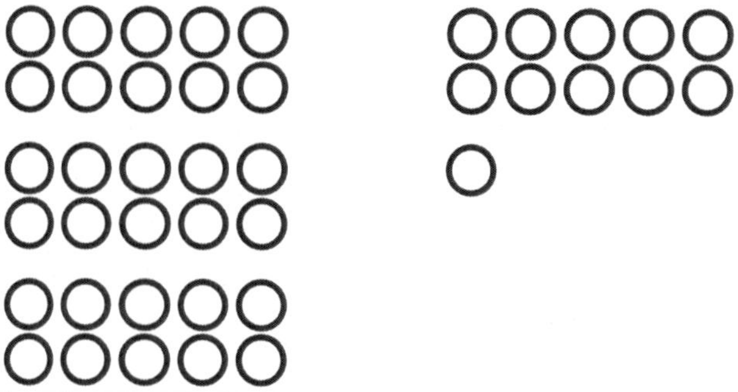

42
forty-two

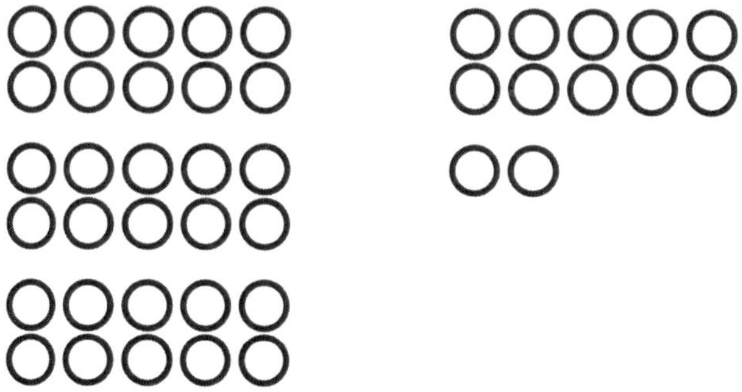

43
forty-three

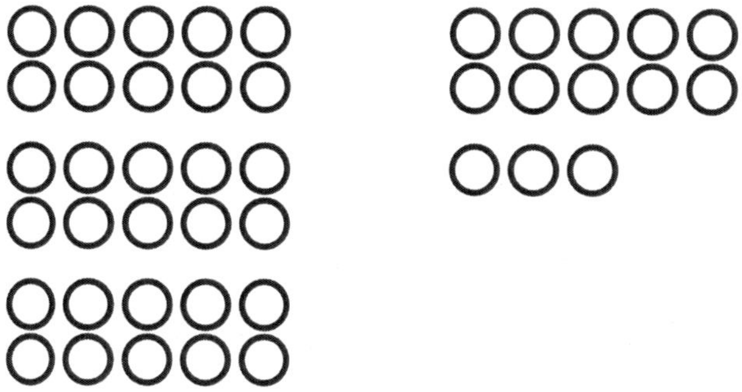

44
forty-four

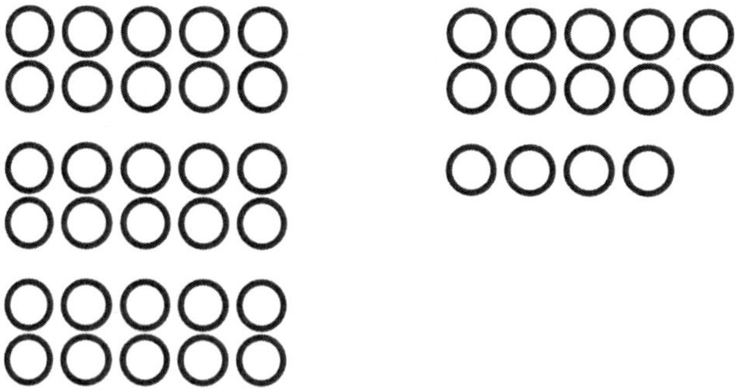

45
forty-five

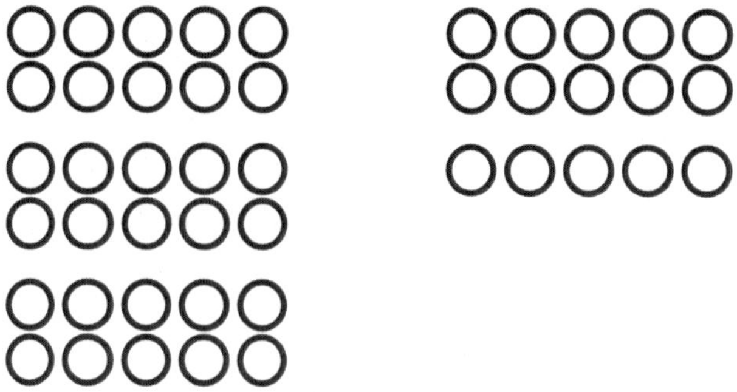

46

forty-six

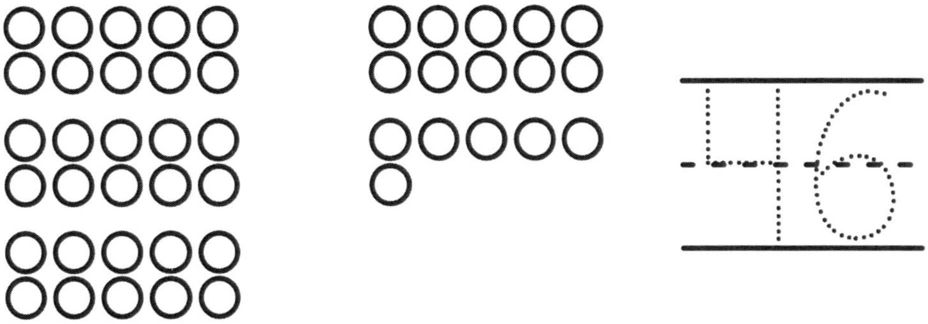

47

forty-seven

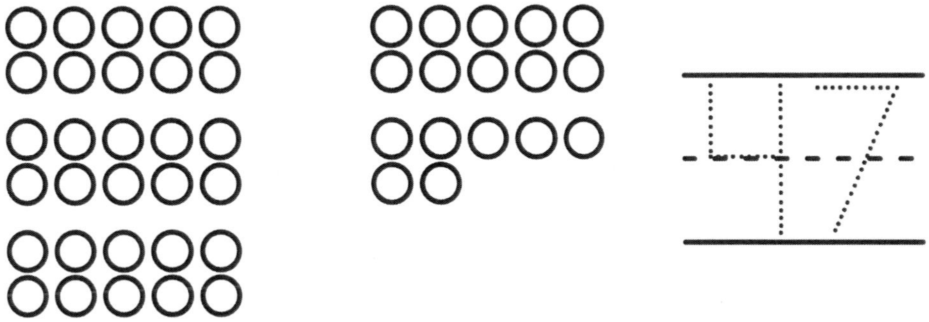

48

forty-eight

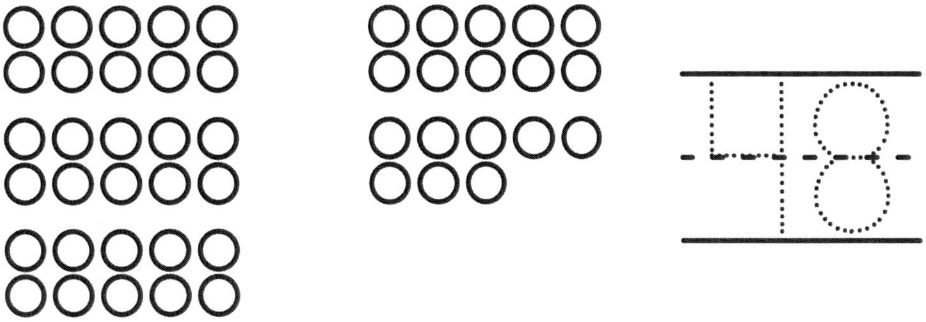

49

forty-nine

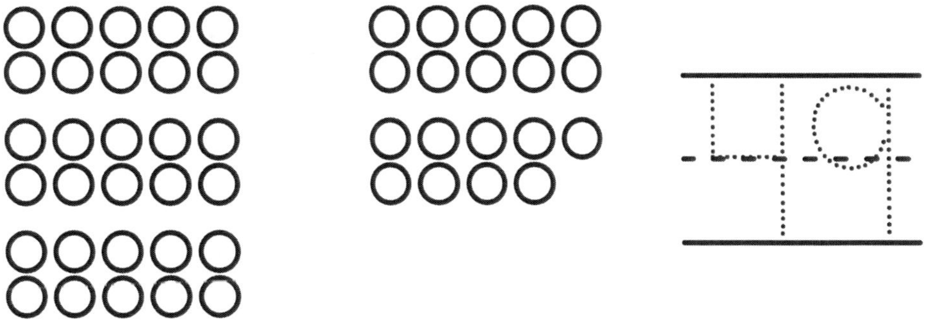

50

fifty

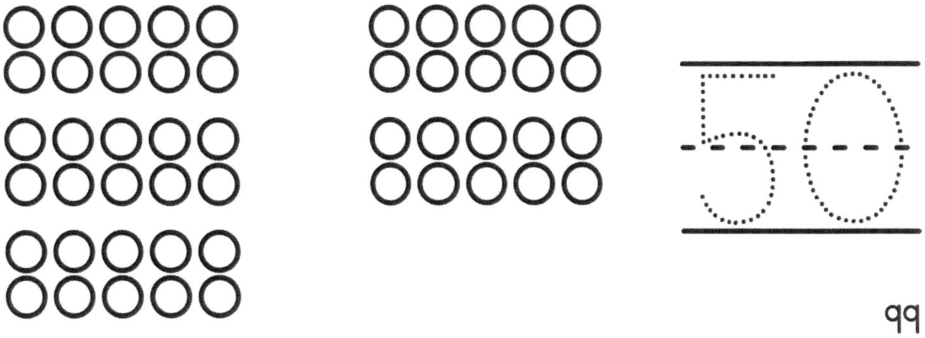

51

fifty-one

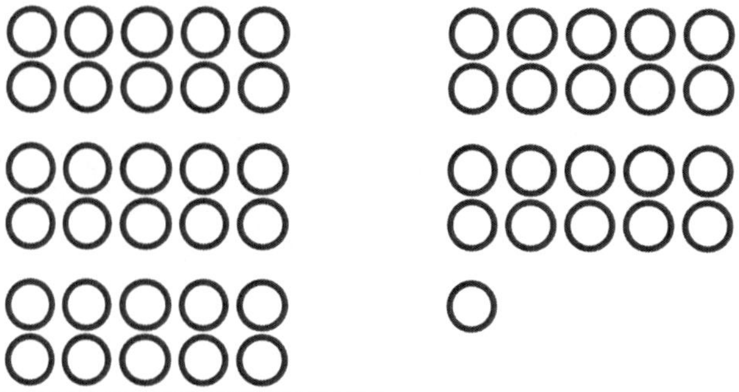

51

52

fifty-two

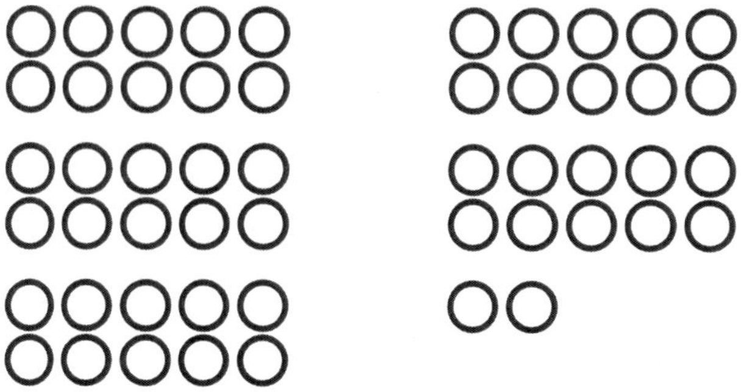

52

53

fifty-three

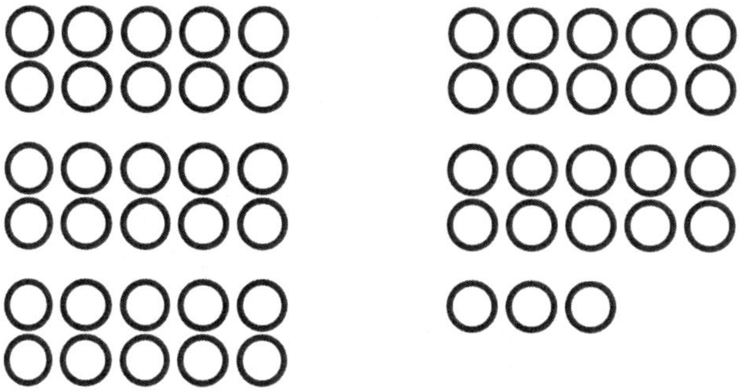

53

54

fifty-four

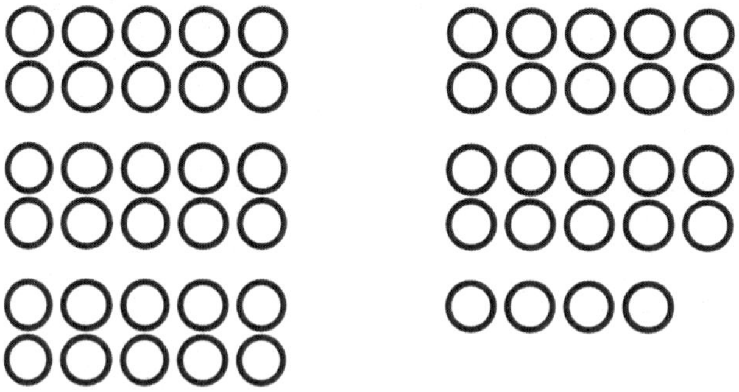

54

55

fifty-five

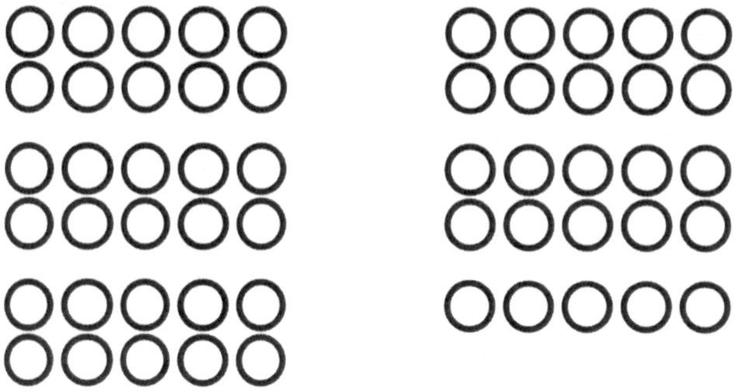

55

56

fifty-six

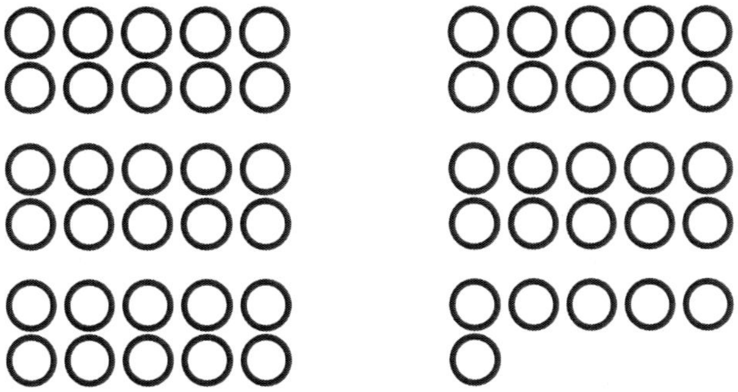

56

57

fifty-seven

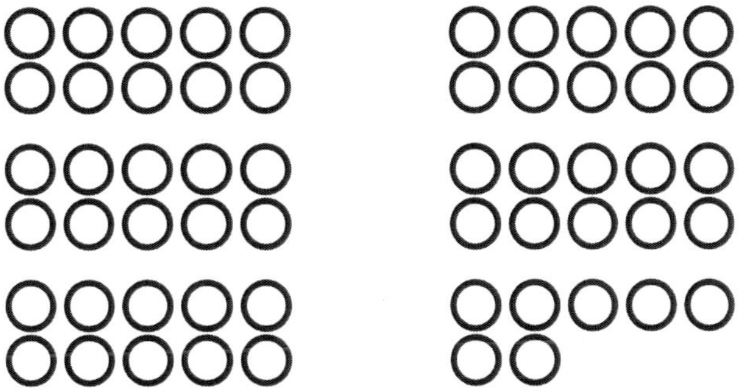

57

58

fifty-eight

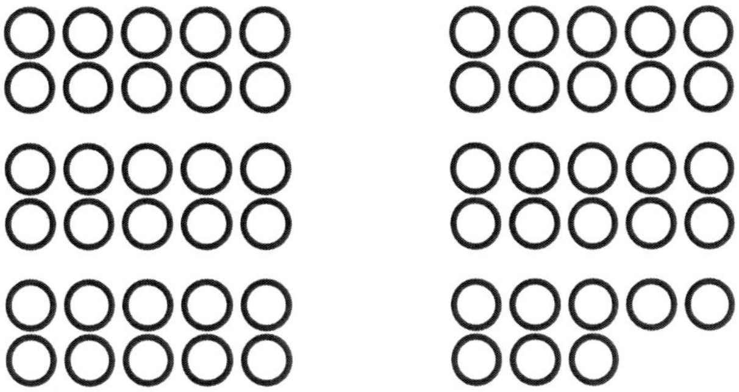

58

59

fifty-nine

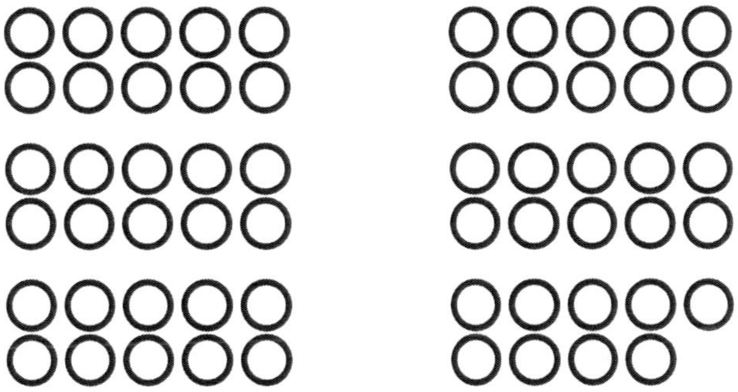

59

60

sixty

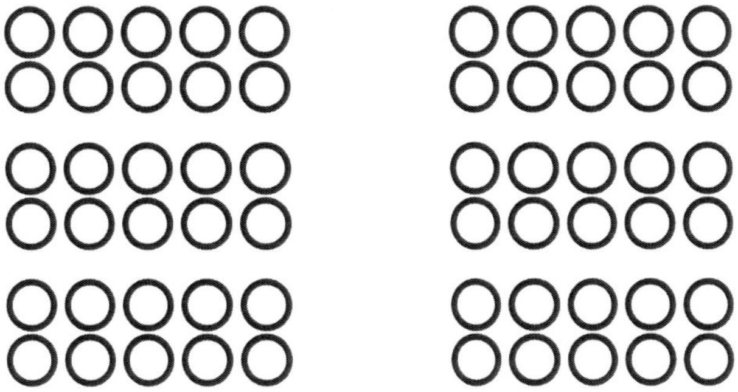

60

61

sixty-one

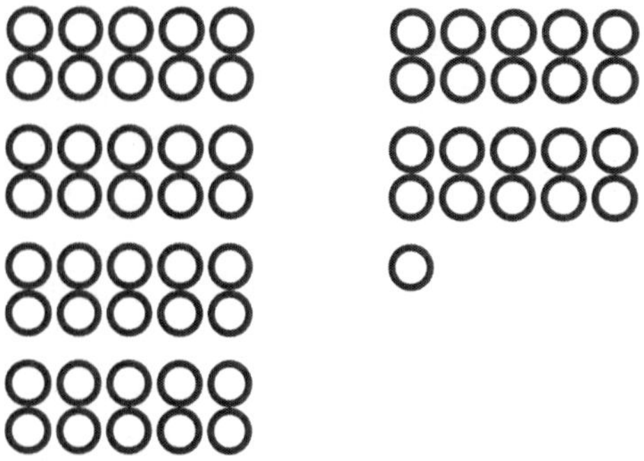

62

sixty-two

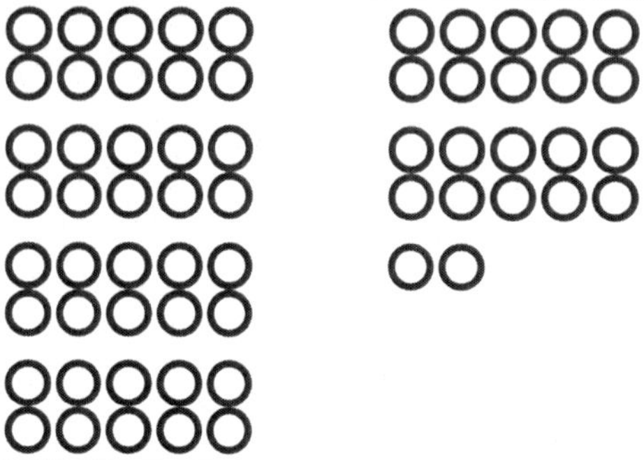

63

sixty-three

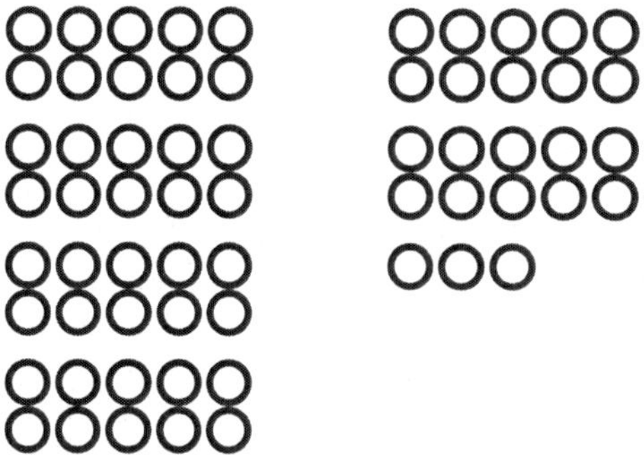

64

sixty-four

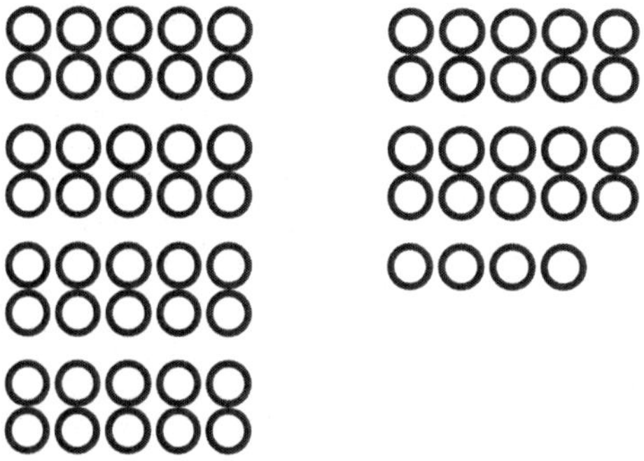

65

sixty-five

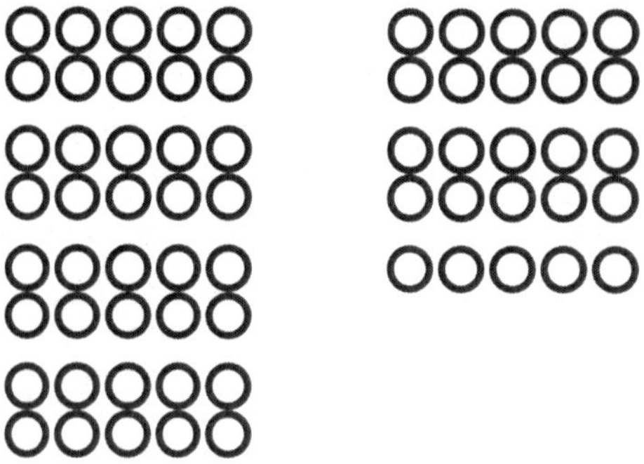

66

sixty-six

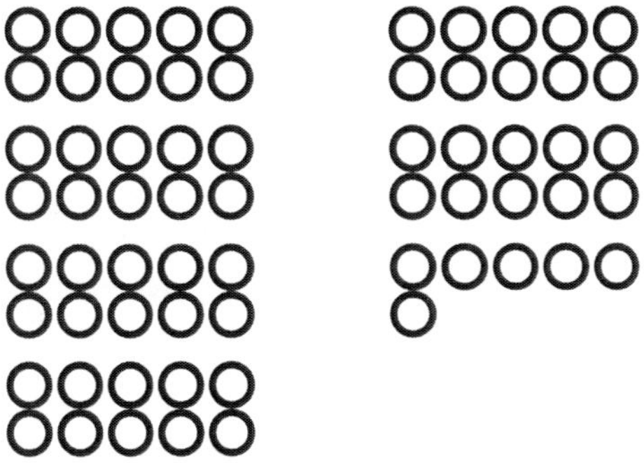

66

67

sixty-seven

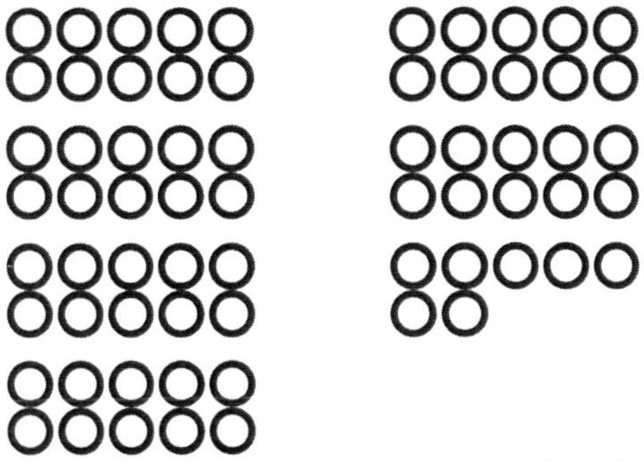

67

68

sixty-eight

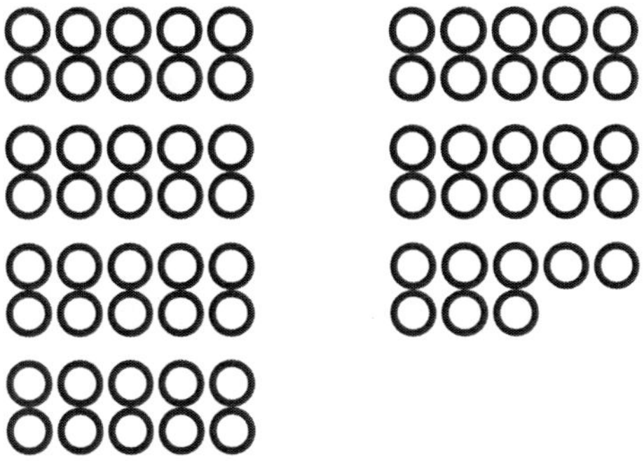

68

69

sixty-nine

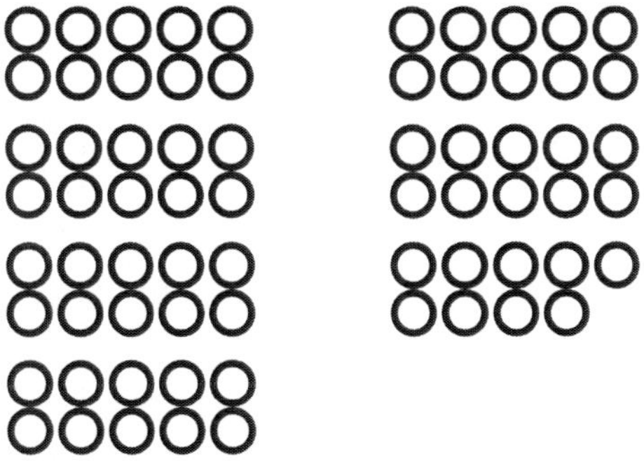

69

70

seventy

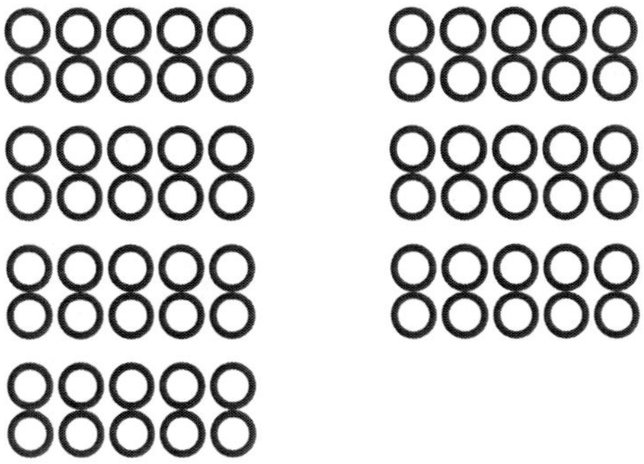

70

71

seventy-one

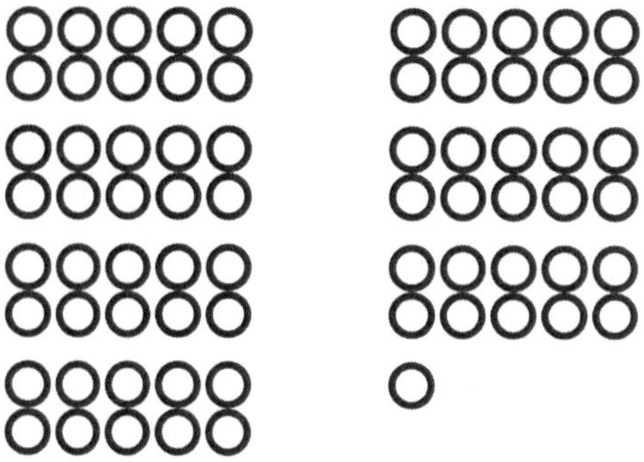

71

72

seventy-two

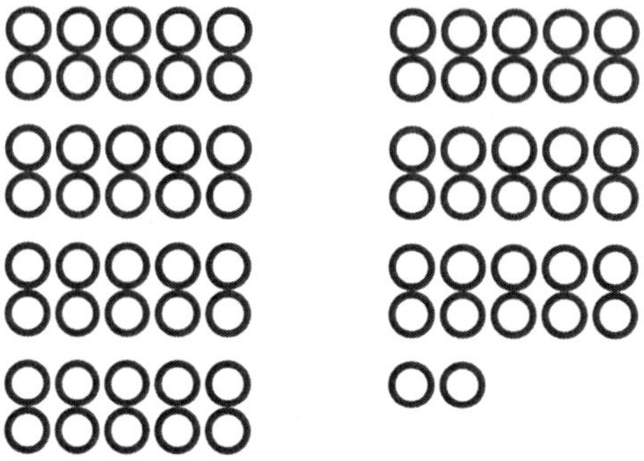

72

73

seventy-three

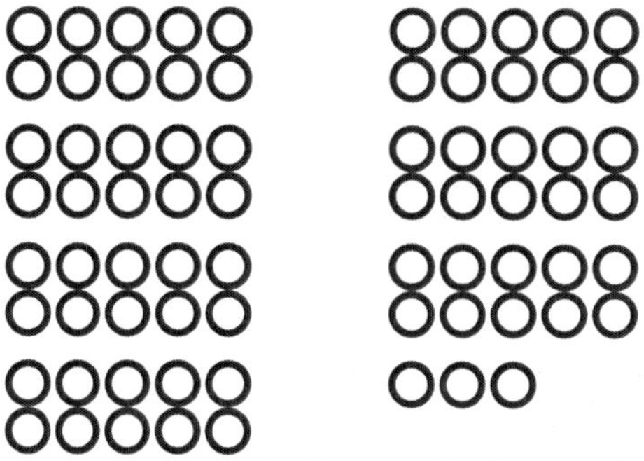

73

74

seventy-four

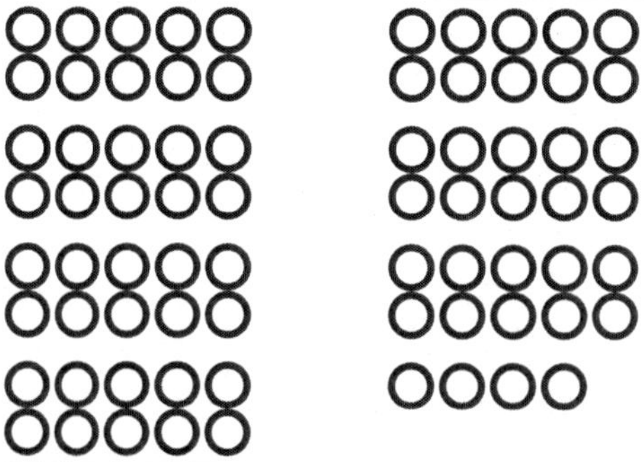

74

75

seventy-five

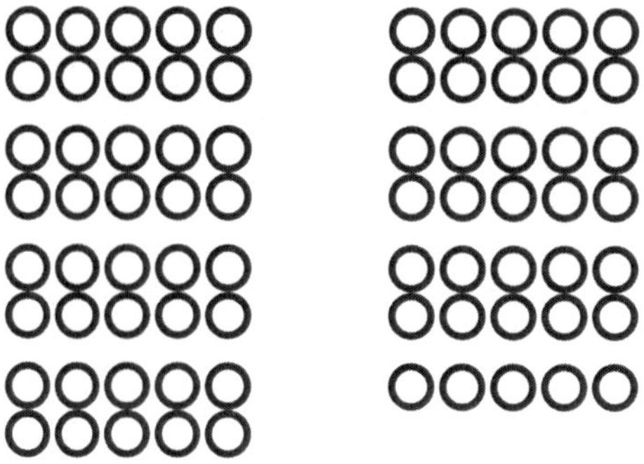

75

76

seventy-six

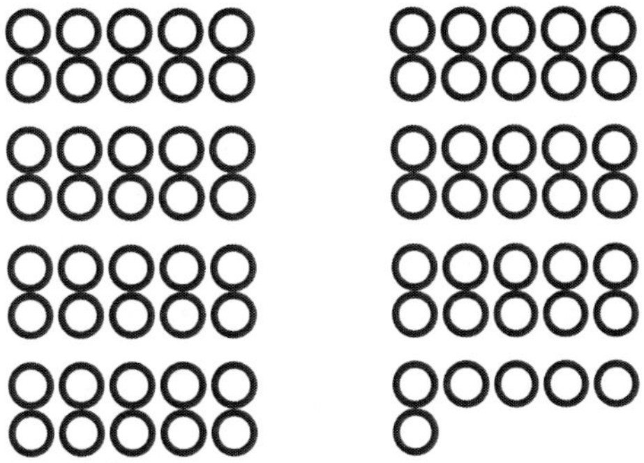

77

seventy-seven

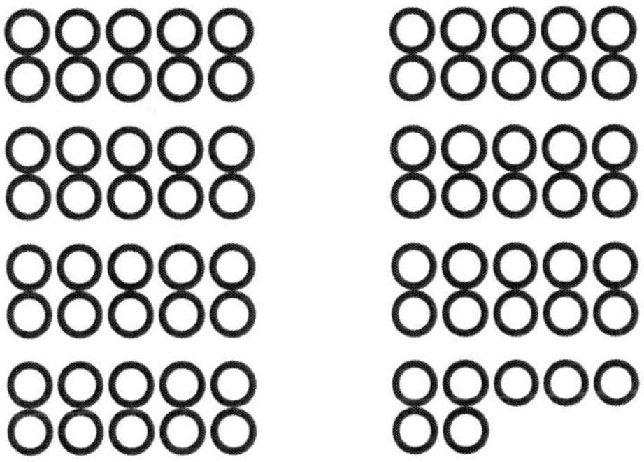

78

seventy-eight

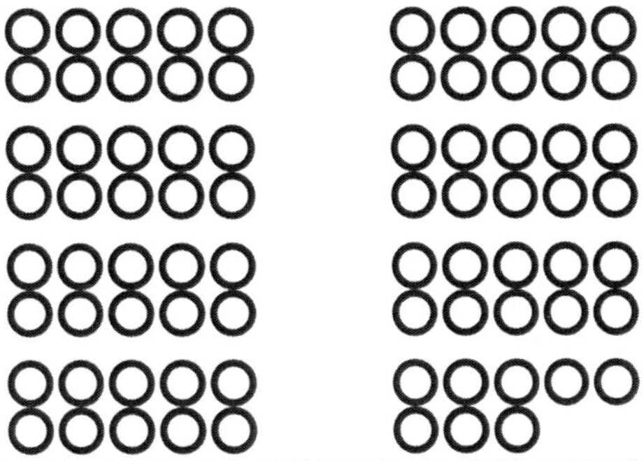

79

seventy-nine

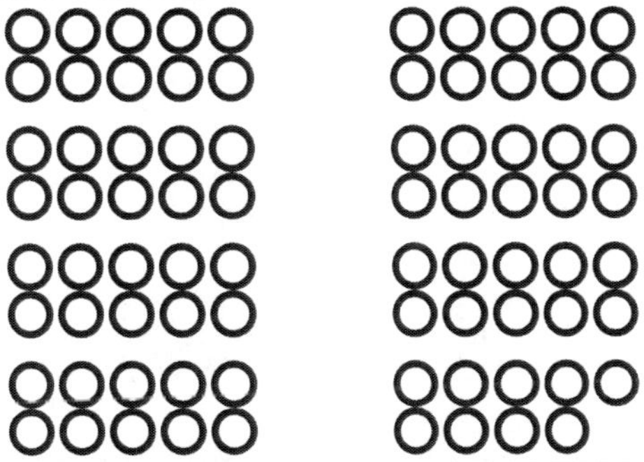

80

eighty

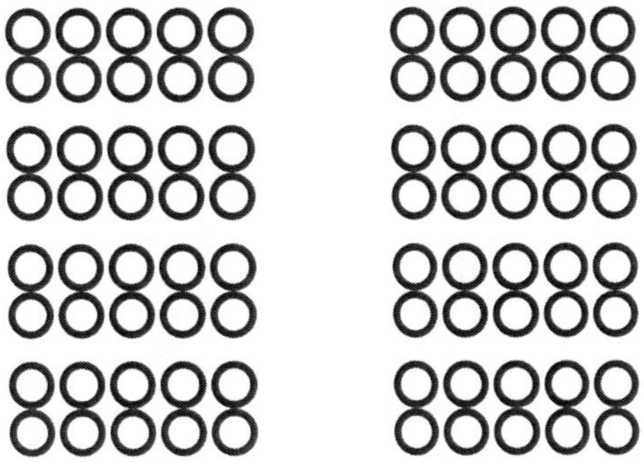

81

eighty-one

82

eighty-two

83

eighty-three

84

eighty-four

85

eighty-five

86

eighty-six

86

87

eighty-seven

87

88

eighty-eight

88

89

eighty-nine

89

90

ninety

90

91

ninety-one

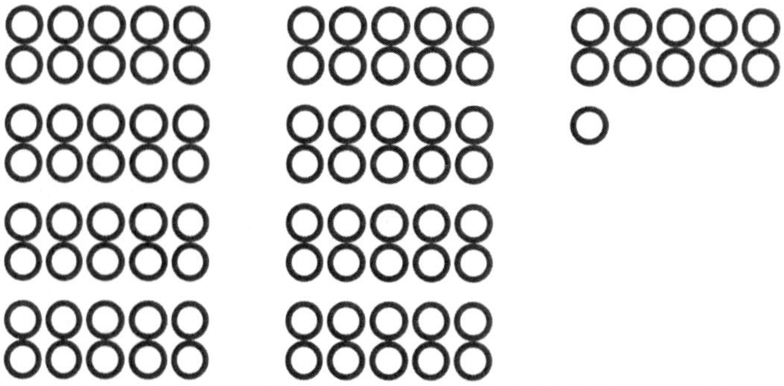

91

92

ninety-two

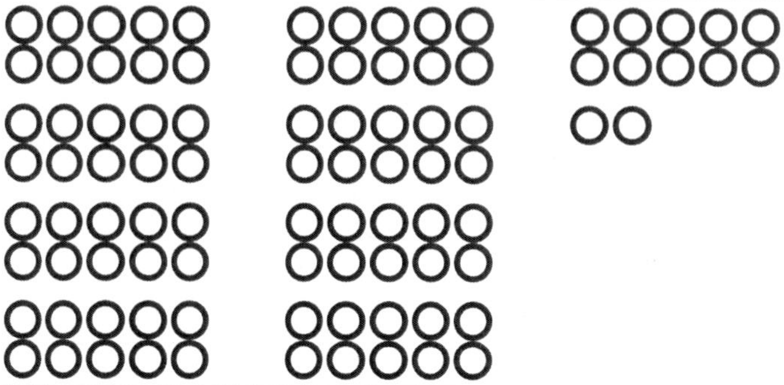

92

93

ninety-three

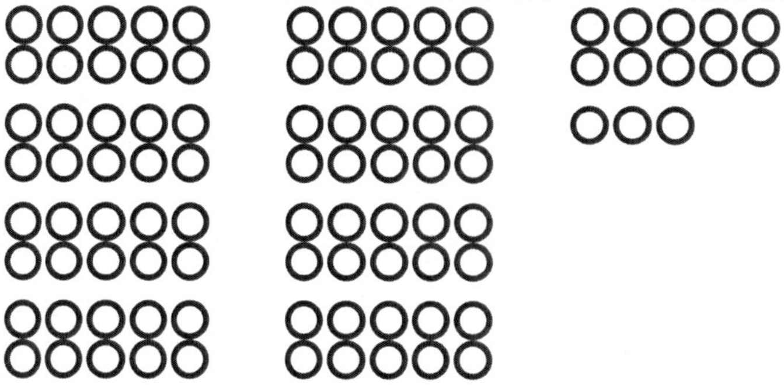

93

94

ninety-four

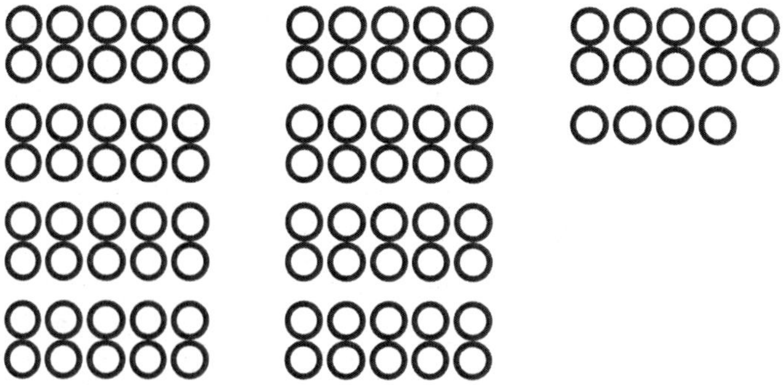

94

95

ninety-five

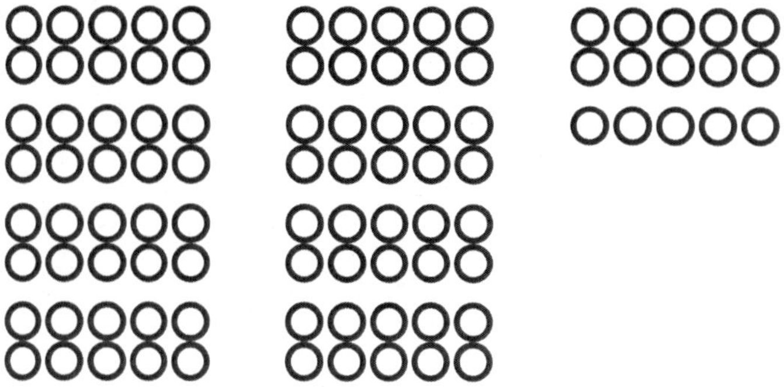

95

96

ninety-six

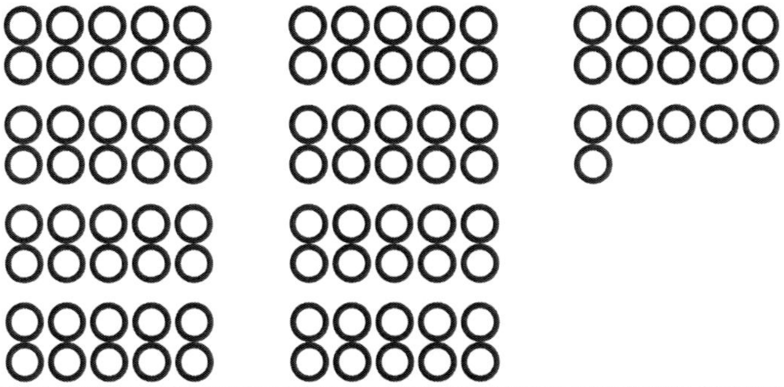

96

97

ninety-seven

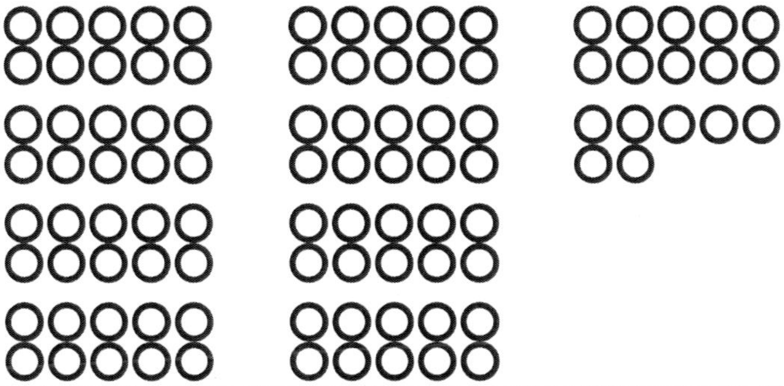

97

98

ninety-eight

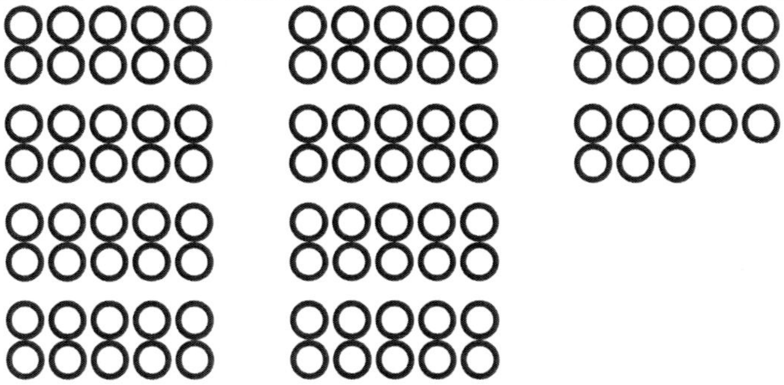

98

99

ninety-nine

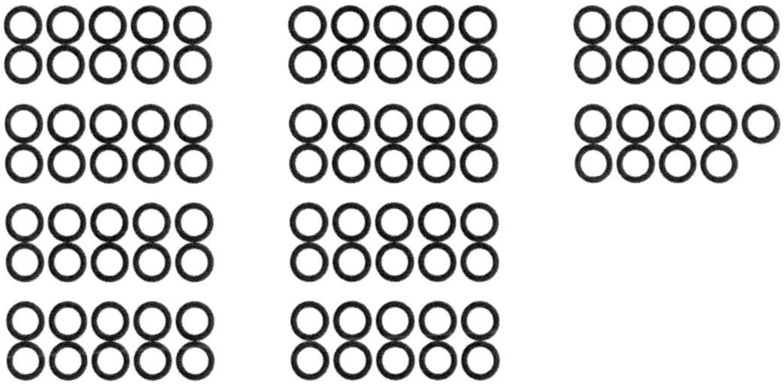

99

100

one hundred

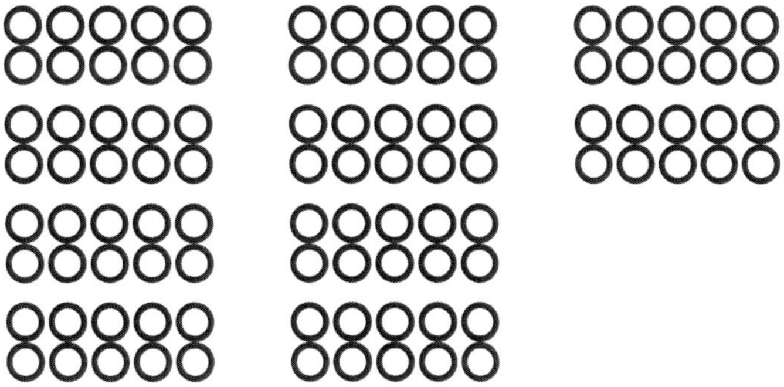

100

TRACE AND WRITE

21

22

23

24

25

TRACE AND WRITE

26

27

28

29

30

TRACE AND WRITE

31

32

33

34

35

TRACE AND WRITE

36

37

38

39

40

TRACE AND WRITE

41

42

43

44

45

TRACE AND WRITE

46

47

48

49

50

TRACE AND WRITE

51

52

53

54

55

TRACE AND WRITE

56

57

58

59

60

TRACE AND WRITE

61

62

63

64

65

TRACE AND WRITE

66

67

68

69

70

TRACE AND WRITE

71

72

73

74

75

TRACE AND WRITE

76

77

78

79

80

TRACE AND WRITE

81

82

83

84

85

TRACE AND WRITE

86

87

88

89

90

TRACE AND WRITE

91

92

93

94

95

TRACE AND WRITE

96

97

98

99

100

Made in United States
North Haven, CT
05 November 2026

21777123R00070